Pigeon Passion

The Complete Pigeon and Racing Pigeon Guide

The ultimate manual for pigeon fanciers

How to win with homing/racing pigeons using minimum effort with maximum speed

by

Elliott Lang

Published by IMB Publishing

© 2010 IMB Publishing

www.pigeonsandracingpigeons.com

Printed and bound in Great Britain by Lightning Source.

ISBN: 978-0-9566269-0-5

A catalogue record of this book is available at the British
Library.

With thanks to my dad for teaching me all about pigeons.

And thanks to my wife and kids for sticking with me throughout the many hours I spent writing this book.

Contents

Introduction

Humanity's relationship with pigeons has been recorded throughout history, in myth, folklore, fact and legend. Whether it's the Biblical image of Noah being led to shore by a pigeon or the lifesaving messages the birds carried during every war, we all should be humbled by man's true best friend.

From there the practice of pigeons has messengers has evolved into racing, pushing it to become a worldwide sport with large amounts of money being made from race prizes and betting. Pigeons are cultivated all over the world for their beauty, their will to survive, their tenacity, their speed and their endurance.

The pigeon is the only bird that has developed such close links with humanity and been useful in so many ways. From time immemorial, the pigeon has served as symbol, sacrifice, food and messenger. It has also played a role as bait and decoy in falconry and thousands died in nineteenth century shooting matches.

Some of the pigeon's illustrious history has been obscured by the confusion between the terms 'pigeon' and 'dove'. Usually, 'dove' is traditionally used in the contexts of religion, literature and art and 'pigeon' for sport and cooking. Historically, the dove has been associated with motherhood and femininity. The Sumerian goddess Ishtar was often depicted holding a pigeon and the ancient Phoenicians associated Astarte, the goddess of love and fertility, with the dove. The Greek goddess Aphrodite and the Roman goddess Venus were also symbolised by doves.

Both Noah's dove and the New Testament dove of the Holy Spirit are the ancestors of today's urban pigeons. Noah's story is echoed in the Babylonian *Epic of*

1

Gilgamesh that includes a story about a great flood and a pigeon playing the role of a messenger.

Noah's story makes it clear that he was familiar with the bird's homing ability. The symbol of the dove carrying an olive branch and bringing its message of hope and peace still endures today.

The pigeon is descended from the blue rock pigeon, found in the wild everywhere in the world except at the polar icecaps. It makes its home on cliffs, but has always had a tendency to nest around human dwellings. People have bred the common pigeon for almost four thousand years. The earliest records of pigeon keeping date from around 2,600 BC, during the fourth Egyptian dynasty, and pigeons can be seen in a number of paintings and hieroglyphics. The Egyptians would release pigeons in order to announce the rise of a new pharaoh.

At first, pigeons were bred in small dove cotes, and later in large structures called *columbarium*. Hundreds of ancient columbaria have been found in Israel, some large enough to have contained thousand of pigeons.

In the distant past, using pigeons was the fastest way to send messages. Trained by the Egyptians and Persians, messenger pigeons spread across the civilised world. Pigeons were at the heart of a great network of communication that kept rulers in touch with the most remote areas of their domain at a time when a horse and rider would have taken weeks to deliver the same information. Phoenician merchants used to take pigeons on their ships and let them go whenever they needed to release information about their business. China organised a postal system based on the use of messenger pigeons and the Greeks used homing pigeons to send news of Olympic victories. In eighth century France, only the nobles had homing pigeons and the birds were considered a symbol of

Introduction

power and prestige until the French revolution.

One of the earliest tame pigeons belonged to the Greek poet Anacreon in the sixth century BC. He wrote a poem describing how his pigeon carried a love letter for him, drank from his cup, ate from his hand, flew around the house and slept on his lyre. Later, in the first century BC, the Jewish philosopher Philo noted on a visit to Ascalon that the pigeon had become 'very bold and impudent' on the domestic scene.

Many societies saw pigeons as a cheap source of good meat, especially during the winter when larger animals were unavailable as a food source. The Romans force fed squabs to fatten them up and wealthy landowners often had pigeon houses. Pigeons were also used as a source of high grade nitrogen (droppings) for fields.

In the nineteenth century, Julius Reuter founded the news service that still caries his name as a line of pigeon posts.

From the time of the ancient Greeks, armies have carried pigeons ready to send news to headquarters. When Paris was seized during the Franco-Prussian war of 1870-1, hot air balloons were used to carry baskets of homing pigeons and other letters out of the city. The pigeons were used to send messages back and, owing to the advent of micro photography, as many as 30,000 messages could be carried by a single bird. During the four month siege of Paris, four hundred birds delivered nearly 115,000 government messages and about a million private messages.

In 1914 when World War I broke out, the armed forces began to use pigeons in their war communications. At this time, the telegraph was the common method for communicating, but telegraph wires were easily cut in two or tapped into by enemy forces. Portable pigeon lofts accompanied soldiers to the front so that they could send messages almost instantly. The British Intelligence Service used pigeons as a way of maintaining contact with sympathisers and resistance movements in enemy occupied territory and the Germans had photographer pigeons with cameras strapped to their bellies.

Batches of pigeons, with their own body harness and parachute, were jettisoned from airplanes and released by a clockwork mechanism. Although a large number perished, ninety five per cent of the birds released returned with essential messages. Over half a million birds were used by the warring armies as reliable communication.

One of the most famous pigeon stories of this time is that of the lost battalion in France that was saved by a pigeon named *Cher Ami*. The battalion was being shelled and wounded by friendly fire because they advanced too far into enemy territory. Several birds were released and when Cher Ami took flight, the German soldiers fired at the bird and wounded it. It arrived back at the command post twenty five miles away with one eye shot out, a bullet in its breast and the leg that carried the message capsule hanging on only by a tendon. The battalion was later saved and Cher Ami received a Croix de Guerre and was taken back to America where he lived until 1919. He was later mounted and then placed on display in the Smithsonian Institute.

In World War II, pigeons were again used by both sides. The head of the SS, Heinrich Himmler, was also head of the German national pigeon organisation and the Germans had 50,000 birds ready for use when war broke out. Although radio had developed enough by then to carry voice rather than Morse code, pigeons allowed radio silence to be maintained while keeping communication open. They also carried cameras over enemy locations to discover information about troop strength and location. Spies on both sides used pigeons, and the birds often flew across the English Channel between Britain and France. Both the English and Germans developed falcon programmes to intercept birds, but the falcons were just as likely to intercept one of their own pigeons.

The Israeli army used pigeons in 1948 during the war of independence to send and recive messages from the sealed city of Jerusalem.

Introduction

Pigeon racing as a sport may date back as far as 220 AD – if not earlier. During the last hundred and fifty years, the modern racing pigeon has been developed to fly further and faster than ever before. A variety of breeds have been combined in the ongoing search for perfection. The most successful modern racing pigeons were developed in early nineteenth century Belgium. The first long distance pigeon race took place in Belgium in 1818. By 1870 there were 150 racing societies in Belgium and over 10,000 lofts. The hobby began to become popular in England a few years later and after that in the US.

The pigeon isn't always popular, in many cities they are regarded as a nuisance. The world appears split between those who love and those who hate them. Feral city pigeons are the descendants of abandoned dovecotes of the past that have adapted to an urban existence. Civic authorities seek to control their numbers using a variety of methods – pigeon proofing buildings, restricting feeding by the public and erecting pigeon lofts from which eggs are removed to reduce numbers.

In seventeenth century America, people frequently commented on the huge numbers of pigeons. Some of their roosting sites covered an area five miles by twelve with up to ninety nests in a single tree. It's estimated that there were five billion passenger pigeons in North America when the Europeans arrived – around the total number of birds to be found there today.

Although pigeons were eaten and used for shooting practice, it wasn't until large scale commercial hunting began in the mid nineteenth century that mass slaughter really began. The passenger pigeon died out in the wild in Ohio about 1900 and the last survivor died in captivity in 1914.

The pigeon family includes hundreds of breeds and pigeons are found almost everywhere on earth, except for very cold or very dry places. Pigeons are powerful flyers and can reach speeds of 70 km per hour.

Scientists are still discovering more about the pigeon. In 1995, three scientists published an article describing an experiment that showed that pigeons can be trained to tell the difference between paintings by Picasso and Monet. The birds were shown a limited set of paintings by the two artists. When the displayed painting was a Picasso, the pigeon was able to get food by repeatedly pecking. If the shown painting was a Monet then pecking had no effect. After a while, the pigeons would only peck when the painting they were shown was a Picasso. At this point the birds were able to generalise and tell the difference between paintings they had not seen before and even discriminate between cubist and impressionist paintings (cubism and impressionism being the two different painting styles of Picasso and Monet).

Above all, we love and race pigeons for reasons that are hard to explain. It's a mixture of the anticipation after a race, before your birds have returned. It's the joy of finding one of your new youngsters may be an amazing bird. It's the joy of putting in time and effort into something you've built up entirely by yourself and that you're proud of.

You are not alone in your love for pigeon racing; there are a vast number of us who feel like you do. I hope you enjoy reading this book as much as I enjoyed writing it.

1

The Pigeon

There is certainly a right way and a wrong way to handle a pigeon. Holding a bird incorrectly will upset it, cause it to lose feathers and even damage its tail bones. When handling a pigeon:

- NEVER handle a bird with damp or sweaty hands. This moisture will damage its feathers. Always ensure that your hands are dry and clean before handling your pigeons.

- Hold your right hand out, palm up, with your fingers together except for your index finger.

- Slide your hand under the bird's left side so that the pigeon is horizontal against you and facing your left hand.

- Allow the birds legs to slide between your index finger and second finger and carefully hold its legs between them.

- A bird should not be handled more than once a day. Covering the pigeon's head with a lightweight cloth may help to keep it calm. The main thing is not to hold the bird too tightly. Pigeons do not have a diaphragm and can suffocate very quickly.

Signs of a healthy bird

- Droppings should be well formed, without smell, brown/green in colour and firm. Birds with slimy, wet droppings or droppings with worm like particles in them should be either culled or removed from the loft.

- Well developed back and breast muscles with strong feathering all over their body. A healthy pigeon will readily give you their wing and look prepared for flight.

- There should be no yellow, slimy mucous or deposits or grey spots anywhere in the lining or on the jaw.

- Their beaks should have white, powdery skin and the same around their eyes.

- Feathers around their anus should be dry and clean with no discolouration or dirt.

- No sign of flies, mites or other parasites.

- Alert, bright and clear eyes.

- Their feet should be clean and responsive. Only an ill pigeon would allow itself to get dirty feet.

Always quarantine new birds: Many birds carry bad bacteria yet look healthy. All new pigeons should be wormed, treated for canker and coccidia, and placed on tendays of antibiotics (Baytril or Cipro are good) to get rid of salmonella BEFORE they are put in with the other birds.

Do NOT cause your birds any undue stress by over handling, irregular loud noises or constant interference. Stress will definitely affect their performance.

Characteristics of a good pigeon

It is impossible to look at a bird and say without doubt whether it is a great specimen or not. The following information is designed as a good guide to how to spot a pigeon that is definitely worth training. Some fanciers can hold a bird and instinctively feel by its weight, breastbone and body firmness whether it is a possible prize winner, but for those of us who are not so fortunate, I present a rough overall guide. I am not saying that birds that tick all these boxes will win races, nor that birds that do not tick them all will never win; I am saying that from experience these birds have the highest chance of success:

Head

Should be well rounded (but sometimes slightly flat on top) and well developed and large, even slightly bulbous, between the eyes. Overly small heads tend to result in a bird that is not as intelligent as his larger headed peers.

Beak

Solid and strong looking and firmly planted in the head. Size should be in proportion with the rest of the body. Upper mandible should be larger than the lower.

Body

Should be of a sturdy build with flesh feeling hard and not soft to the touch. The back should be broad with strong shoulders while the chest should be deep with enough room for large, well developed lungs.

Rump

Covered completely with thick feathers.

Breastbone

Obviously the wing muscles are connected to the breastbone, so if it is not well formed then the bird will not fly well. The breastbone should therefore be strong and thick to give hold to these muscles. If a pigeon has an arched breastbone then it will never be a good racer and should certainly be avoided.

Wings

Supple when stretched; not too easily giving and not overly hard to open. Primaries will offer great resistance due to the small feathers covering the wings. The larger and stronger these feathers, the better the wings. The tip of the wings should be very close to the end of the tail but wings do not necessarily have to completely cover the rump.

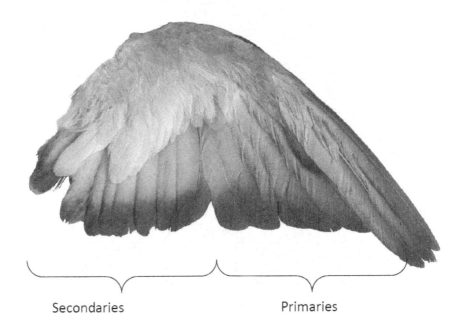

Secondaries   Primaries

Labelled diagram of a bird's wing, indicating primaries and secondaries

Eyes

Time and time again I hear people talking about birds they've spent hundreds or thousands of pounds on because they read its 'eye sign' and it's 'definitely a winner'. It's all nonsense and all tests to try to validate this have failed. Any fancier trying to convince you that it works will be basing what they say around their own hindsight bias.

The Pigeon

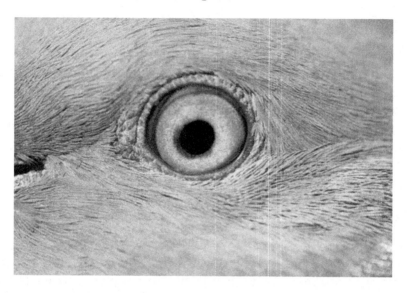

What do I mean?

Well let's say a fancier is a massive believer in a birds potentially being linked solely to the eye sign. He buys one hundred birds all with eyes that tell him they're going to win every race. Out of these he'll have bought ninety poor racers, nine average racers and one amazing bird. He will forget the ninety nine birds that didn't live up to their eye signs and will, instead, be amazed at his singular pigeon that conformed to his pre-conceived belief and will now believe in eye signs more strongly because he was right 1% of the time.

It's like horoscopes. I could read mine every day for a year and never have one accurate prediction. Then, just once, the horoscope roughly matches something that happens to me that day. Was this horoscope inexplicably predicting the future? No, it was just a statistical likelihood that at some point, if I read enough horoscopes, one of them would match something that actually occurred.

Plumage

Colour is unimportant but the quality of the quills matters. Quills should be flexible whilst the feathers are silky and smooth. Good plumage has a shine on the bird's neck and throat. The surface area of the wing should look like a marbled flat and hard surface. A healthy bird will deposit a white bloom on contact. If you find a bird with split quills then it needn't be a massive cause for concern; just don't breed the bird with another with the same peculiarity to avoid a genetic repeat.

Tail

Slender and not overly long. At rest it should have a width of one feather. As with all the previous tips, this is not definitive, a wider tail is not necessarily a disadvantage.

Feet

Clean, sturdy looking and nimble. If your bird is ill you can almost always spot it first in the feet: healthy birds will always keep their feet clean whilst sick birds can let their feet get dirty.

2

Pigeon Breeds

The domestic pigeon, (columba livia) has been bred for over three thousand years. It is descended from the rock dove or rock pigeon and has many close relatives, the closest being the stock dove (columba oenas).

Today, a wide variety of forms of the domestic pigeon exist. Some are bred for flying abilities, including tumbling or rolling (somersaulting in the air), diving and homing. Others are bred for their voice (trumpeters and laughers), beauty, feather ornaments or uniqueness. In some cases a pigeon type is developed that includes more than one distinct characteristic.

All domestic pigeons are edible and some are still bred for food. These birds (squabbing pigeons) are normally much bigger than most show or flying breeds. The largest domestic breed (runt or Roman pigeon) may reach weights of up to 1.5-2.25 kg. The basic pattern of the common pigeon in its domestic form is 'blue bar'. This is a mix of the pigment melanin, and produces dove grey and black. A green/purple iridescence caused by the structure of the feather can be seen on the neck and chest front. The blue bar pattern gets its name from two black bars which cross the wing towards the back.

Another very common pattern is check (chequer, checker), which is similar to the blue bar pattern but also has many little black marks across the wing shield. It can range from light to very dark. The barless pattern, which has no black bars or checks across the wingshield, is rare and occurs mainly in breeds of Central European ancestry.

Each of these patterns may also be found in the other two pigment series of ash red and brown. Other common colours include white, black, recessive red, pied, splash and grizzle. Birds are normally designated by both their colour and pattern. The names used to describe the different combinations can vary from place to place.

The main categories of pigeons are:

Performance pigeons

Racing homers

- Birmingham rollers and flying tumblers - Small pigeons that are flown over the home loft, for a period of time. While flying, they exhibit a series of backward flips.

- Flying tipplers - Pigeons flown over the home loft, for an extended period of time. The idea of tippler competition is for the birds to fly as long as possible.

- Parlour tumblers and rollers - Pigeons that have lost the ability to fly and that roll on the ground or do a series of flips just above the ground.

Show pigeons

- Pigeons that are strictly for the show pen. Judged to a strict standard of excellence, this category of pigeons is rarely released to fly.

Utility pigeons

- Pigeons bred for their size. This is a category of pigeons bred for eating.

Pigeon breed examples

Antwerp - These large pigeons are short, medium or long faced with a shapely, broad breast, deep chest, a long straight back and broad shoulders with tail feathers that don't touch the ground. They come in many colours with red eyes, a black short beak and medium length legs with crimson feet.

Aachen Lacquer Shield Owl - A breed of fancy pigeon.

African Owl – A small breed of fancy pigeon with a short beak.

Altenburger Trumpeter - One of the several breeds of trumpeter pigeons, which are known for their vocal cooing which sounds similar to laughter or trumpeting.

American Giant Runt - Known for its large size and suitability for squab production.

American Show Racer - Known as the 'Bird of Dignity', this breed began in the early 1950s. Emphasis is placed on 'station' (which includes an upright posture) and a powerful head. The bird should be very smooth feathered. The breed is popular at shows.

Antwerp Smerle - One of the breeds used in the development of the *Racing Homer*.

Arabian Trumpeter – A trumpeter pigeon.

Archangel - A fancy pigeon known for the metallic sheen of its feathers and unusual appearance. It has unfeathered legs and dark orange eyes and may be crested. The body of the bird is bronze or gold with wings that are either black, white or blue. It is also known as the *Gimpel* (German for  European Bullfinch). In the US, all colour forms are referred to as *Archangels*. In the UK, only the black and copper coloured birds are called Archangels. This is a very old breed, and has a slender neck and body with beautiful plumage.

Armenian Tumbler - One of a breed of tumblers that originated in Armenia, it is bred for colour and flight. It has distinct markings with many having a black neck with a black tail, although others have a yellow tail and yellow neck (known as a bellneck). Their feathers are clear smooth but strong, they might also have feathers on their feet. It has a round head and can also be fully crested or non-crested. It usually has a medium sized light pink coloured beak or a black beak. Its eyes can be pearl, orange, and yellow, and they can have two eye colours - one for each eye.

Australian Performing Tumbler – This was developed in Australia and has been a popular flying variety in the past due to its spinning/rolling action. Nowadays, it is maintained mainly for exhibition purposes. The APT is medium faced, pearl eyed, clean legged with a rounded head, short body and it comes in a selection of classic tumbler colours such as recessive red, kite and almond.

Australian Saddleback Tumbler - First recognised as a breed in 1917, it is usually muffed (feather legged), although a clean legged variety also exists. Its name derives from a distinctive saddle shaped marking on the back. The marking on the head can be either a stripe (most common) or spot. The most common colours are black, red, yellow, blue, silver, almond and andalusian.

 *Barb* - Also known as an *English barb* or *Shakespeare*. It has a noticeably short beak and a rounded head. Sometimes, they are referred to as *whiskered pigeons*. Their profuse feathers and frilled feather pants make them stand above a common pigeon. The barb is a medium sized pigeon with short, thick curved beak, long body and a short neck. The knobby flesh around the eyes called the cere is coral red and the eyes are white with black pupils. Its colours are white, yellow, dun, black and red.

Birmingham Roller – It originated in Birmingham where it was developed for its ability to do rapid backward somersaults while flying. Show Rollers are larger than the flying variety.

Bokhara Trumpeter - Known for its long muffed (feathered) feet and double crest, which completely obscures the bird's eyes.

British Show Racer - It was developed as an exhibition breed in Britain from local stocks of racing pigeons.

Budapest Highflier (Poltli) - This breed is most popular in its native Hungary.

Budapest Short Face Tumbler - Created in Budapest, Hungary.

Carneau - A utility pigeon from Belgium and France, it is a show bird with a short, compact, heavy set broad breast. It has large eyes with a smooth, flesh coloured cere, a beak of medium length and is stout with a

V-shaped wattle. Its colours are yellow, red, white, black and dun and the plumage is close fitting. The breed is known for its large size and suitability for squab production.

Carrier - Originating in Persia, this breed was thought of for years as the 'King of Pigeons'. It was bred to carry messages and raised because of its homing ability but now is just a fancy variety. It has a large body, enormous wattles around the beak and close fitting plumage and is also tall. Its colours are red, yellow, white, dun, blue and black.

Chinese Nasal Tuft - One of the oldest and best known pigeon breeds from China. A medium sized pigeon bred in black, red, and blue and their dilutes.

Chinese Owl - This breed is known for its small size and profuse frilled feathers.

Cumulet - Originating from France, this breed is a medium sized pigeon, has a full chest and has a well proportioned body with long wings and short legs. The colour is usually only white but some have red flecks on the neck and head.

Damascene – This breed is thought to have originated in Damascus, Syria.

Danish Suabian - It comes in silver, blue, mealy, black, red and yellow and is either plain head or peak crested.

Danzig Highflier - Originated in Danzig (now Gdansk) in 1807.

Domestic Show Flight – A relatively recent American creation, which was developed in the state of New York.

Donek - The name *Donek* is of Turkish origin and means, 'falling down from the sky'. The breed is known for its aerial acrobatics (spiral diving) and is more of a performance breed rather than a purely fancy variety.

 *Dragoon* - A very old breed of British origin, the Dragoon was one of the breeds used in the development of the racing homer. It has a wedge-shaped head, with a short blunt beak and peg-shaped wattle. Its colours are a dark reddish purple in blues, checkers, grizzles and silvers. This pigeon is poised, has a wedge shaped body, short legs and a thick short neck. It is a heavy pigeon.

Dresden Trumpeter – A double crested bird with a white wing shield, dark orange eyes, broad wings and a long tail.

Egyptian Swift - This breed is known for its long wings and tail and its short beak.

English Carrier - Developed in England from the *Persian Wattle Pigeon*, the largest of the flying pigeon breeds, the *Old English Carrier*, was originally used for sending messages. Today, it's a show pigeon, having been supplanted by the Racing Homer. It has a long, slender body and thick wattle.

English Long Face Tumbler - This breed is available in both clean legged and muffed (feathered legs) varieties. Due to its short beak it requires foster parents to raise its young.

English Pouter - A very old breed, known and recorded from the 1730s.

English Short Faced Tumbler - One of the oldest breeds.

English Trumpeter - One of the most popular breeds in the USA, it combines a tuft, crest and large muffs on its feet. It comes in a number of colours.

Fantail (English Fantail) - The fantail originated in India and is known as the *broad tailed shaker*. It has a small head, slender neck, fan shaped tail with a small body and a chest carried  upright higher than its head which rests on the cushion formed by the tail feathers. The most popular colour is white, but it also comes in yellow, red, blue, black, silver, dun, checker and saddle. Fantails are often used in the training of racing pigeons and tipplers. They are used as *droppers* and placed on the loft landing board as a signal to the flying birds to come in and be fed. The Indian fantail was introduced to the US in 1926 when a shipment of four pythons was on its way to the San Diego Zoo. To ensure the reptiles didn't go hungry, Indian fantail pigeons, found exclusively in India until then, were given to the pythons as snake feed. By the time the ship reached California, only two had survived. The zoo keepers were so taken by their distinct looks that they decided to keep, and later breed and develop them.

Felegyhazer Tumbler - The name is short for Kiskunfelegyhaza, a town in the Hungarian lowlands.

Florentine - Originating from Italy, this is also called a *hen pigeon* as its shape is similar to that of a hen. It is a large pigeon with a coloured head, coloured wing covers and tail and white wings. The colours include red, yellow, blue, black and black bars.

French Mondaine - Originally developed in France as a utility pigeon, this bird has a broad back and a long breast.

Exhibition Homer - Originated in England. It is lighter than the Show Homer and has a straight stout beak.

Frillback - The breed is known for the *frill* or *curls* on the wing shield feathers. The feather curl should also be present at the ends of the foot feathers.

Galați roller - Created in Romania, near Galați.

Genuine Homer - Originating in England, this breed is the exhibition counterpart of the Racing Homer.

German Beauty Homer - Originally from Germany and developed around one hundred years ago.

German Coloured Tail Owl (Königsberger Farbenköpfe) - Developed in the mid eighteenth century in Königsberg, East Prussia, from which it gets its German name. This breed of pigeon is well known in Russia because Königsberg is now a part of Russia (now known as Kaliningrad). It has a black, red, white, yellow and blue coat. Its rudder and head can be of any colour. It has perfect body posture, a lordly air and high carriage.

German Nun (Deutsches Nönnchen) – This breed originated in the early seventeenth century and is widespread in Russia, where it is known as a *Cross Nun*. The breed got its name *nun* from a coloured cap on its head, and *cross*  from the coloured tail and coloured primary wing feathers, which resemble a cross during the pigeon's flight. The rest of the feathering is white. The birds have a specific feathering pattern, where the coloured feathering can be black, red, yellow, coffee brown, ash grey, light blue or silver. It's of medium size with a round, narrow dry elongated high forehead. A characteristic feature of the breed is a high, tight, conchiform forelock, which reaches to the crest on the back, and curls on the end of the forelock which descends to the ears. The distinctive flight pattern of German Nuns is a low, circular soar, which usually doesn't last long. If Nuns see an alien pigeon when sitting on the roof, they take off and clap their wings to 'invite' that pigeon to their loft.

Ghent Cropper – Its feather muffs extend to its feet and this pigeon is often noticed in soft colours of light brown, grey, white or cream.

Giant Homer - Originated in the US for its size and squab producing ability. It comes in many colours with blue checks and silver being the most popular.

Show Homer - A large pigeon with a head which forms a long unbroken, well arched curve from the tip of the beak to the back of the head. It is bred in many colours and originates from England.

Helmet – This breed has a medium face, short face, plain head and crested varieties.

Holle Cropper (*Amsterdam Balloon Cropper*) - This breed was developed in Holland.

Hungarian - Originally from Austria, this is a cross between the Florentine, the Swallow and another pigeon. It is a large, handsome hen type pigeon and is bred in many colours.

 *Ice Pigeon* - Known and named for its ice blue colour. It originated in Germany, and there are two types – clean legged and muffed. Although they are all of the same colour, some have black or white bars.

Indian Gola - A small pigeon available in all colours. It has red eyes.

Iranian Highflying Tumbler - Bred in Iran for endurance flying competitions, this bird's tumbling is just a flip, occasionally hovering before it does the flip. The best birds tend to rise above the rest of the kit to show off their talents. The flying characteristic of the Iranian Highflier is that of a soaring/hovering bird, with a slower wing beat than most flying breeds of pigeons. It is an endurance flyer that gains altitude quickly and comes in a variety of patterns and colours.

Jacobin - A feather hooded pigeon with the peculiar habit of keeping its face covered. Its hood resembles those worn by the Jacobin order of monks. It originated in Cyprus and comes in yellow, white, black, blue, silver and red.

Kiev Tumbler (Kijivskij svitlij, Kievskie svetlie) - A flying breed that flies at a medium height. It is a small, slender and gentle pigeon.

King - It originated in the United States and is a cross of the Swiss Mondaine, Dragoon, Duchess and Florentine. It is a medium sized pigeon weighing with a chunky build and a large well rounded head. The breed was developed during the 1890s and it comes in red, yellow, dun, white, blue and silver.

Komorner Tumbler - The breed has American and European varieties that are recognised as separate breeds. Originally bred for acrobatic flying, Komorners are seldom free flown today. The breed originated in the  eighteenth century in Komárno in the Austrian Empire (now on the Slovak-Hungarian borders). It was imported into the US in the late 1920s. Komorners are small, slim pigeons and typically sport a magpied pattern with colours in black, blue, red, silver, yellow and dun. It is adorned with a crest extending from ear to ear terminating in rosettes.

 *Lahore* - Named after the Pakistani city, this breed was imported into Germany around 1880 and became popular at the beginning of the 1960s. Usually found in the area of Iran, this breed was once bred for meat, but today is raised for its beautiful plumage and colourful patterns. It is a large pigeon with unusual markings - the base colour is white, with a secondary colour beginning at the juncture of beak and wattle and spreading in an arc over the eyes and across the back and wings. The rump and tail are also white, and the neck is heavily feathered leading to a full, broad chest. Lahore are bred in many different colours, including blue bar, checkered, red, blue, brown and black.

Lark – Originally from Germany, it is a large pigeon, broad breasted and long bodied and has two varieties, the *Coburg Lark* and the *Buremberg Lark*.

Lucerne Gold Collar – Originally from Lucerne, Switzerland, it is a variety of the *Swiss Lucerne* peak crested pigeon.

Magpie - Originally from in Germany, this is a small, graceful, streamlined pigeon with a shallow body, snake neck and small head. It has a white body, wings, shoulders, legs, coloured head, neck, chest, back tail and rump and comes in blue, silver, black, dun, cream, yellow and red. The original Magpie was one of the old tumbler varieties.

Maltese- Originally from Germany and Austria, this large bird has a hen shaped body, long neck, long, straight legs and straight tail feathers. The colours are black, silver, blue, red, white, dun and yellow.

Modena – Originally from Italy, it comes in two basic patterns in about 150 different colours. It is heavy set but also graceful. There are two main varieties. *Gazzi* has a pied marking with the head and portion of the throat, the wings and the tail coloured and the rest of the bird white. *Schiettis* are non pied.

Norwich Cropper – It is thought to have been developed from a breed called the *Oploper* and is of Dutch origin.

Old German Cropper - A rare species the head of which is often difficult to locate.

Old German Owl - Originator of the short faced German Shield Owls, this was the first breed in Germany to be called *Mövchen* (little gull) due to its resemblance to the silver gull. It was formally recognised in Germany in 1956,  but the first official standard was not adopted in Europe until 1960. It has a nearly round, broad head with a well arched forehead and a small full shell crest, closing with rosettes. It comes in blue, ash red, recessive red, brown, spread, checks, and bars in black, red, brown and white and dilutes. The body colour is pure white.

Oriental Frill (*Hünkari*) – This breed originated in Turkey and has a frill of feathers on the breast. It has a short beak and peaked crests that rise to the highest point of the head. The varieties include *Satinettes, Blondinettes, Turbiteens,* and *Oriental Turbits*. This bird is originally a Turkish pigeon breed specially bred for the Ottoman Sultans in the Manisa Palace, Turkey. It was first imported to England 1864. It is a small to medium sized pigeon and its typical characteristics include a breast frill, peak crest, grouse muffs and a medium-short thick beak. Satinettes are shield marked/tail marked birds with white bars or laces on their shield and moon spots or laces on their tail. Blondinettes are whole coloured birds which also possess white bars or lacing on the shields and moon spots or lacing on the tail. Some varieties have the lacing extending over most of the body.

Oriental Roller - The most important hallmark of the Oriental Roller is its flying style. They show a variety of different figures in the air, including single and double somersaults, rolling, rotation with open wings, nose dives, sudden change of direction during flight and, very rarely, axial turns.

Owl – This breed originated in Asia and has clean legs, plain heads, and frills of feathers on the breast. The colours are blue, black, silver, white, yellow and red.

Parlour Roller - The breed is known for its unique performance of turning somersaults on the ground. Adult parlour rollers seem to lack the capability of flight.

Polish helmet (Kryska Polska) – This is distinctive on account of its muffs and is coloured only on the top half of its head (helmet) and on its tail.

Pouter - There are many varieties of pouter with little in common except for the nature of the crop. The breed originated in Saxony and Thuringia in the early 1800s.

Runt - The largest of all domestic pigeons, it comes in blue, silver, red, yellow and black.

Scandaroon (Nuremberg Baghdad) - It originated in Baghdad and resembles the Carrier in bearing, shape and size. It has a long more curved beak and comes in blue, white,  black, red and yellow with markings similar to the magpie.

Strasser - Originated in Austria as a utility pigeon, this bird has a coloured head, neck, wings, and tail with coloured feathers on the back. The rest of the body is white. The colours include, black lace, lark colours, blue,

blue barred, black or white barred, blue checkered, red or yellow.

Saxon Fairy Swallow - The body of this pigeon breed is enrobed with coloured muffs.

Saxon Field Pigeons - Saxon Field Pigeons are a combination of various varieties which are basically the same type differing only in colour and markings. Some varieties include: *Saxon Breast Pigeon, Saxon Monk* (pictured), *Saxon Priest, Saxon Reversewing, Saxon Shield, Saxon Spot* and *Saxon Whitetail*. The Saxons are all muff legged.

Saxon Monk - The Saxon Monk comes in five colours including blue, black, red, yellow and silver with white bars or spangles.

Saxon Shield - Saxon Shield comes in black, blue, red and yellow with white bars or spangles. The blue has black bars or is barless.

Saxon Spot - Saxon Spot comes in four colour varieties: black, blue, red and yellow.

Schmalkender Moorhead - Originally from Germany, it has a high, arched head and a long tail carried closed. The head, bib, tail, rump, and wedge under tail are coloured and the remaining feathers white. It comes in black, blue, red and yellow.

Serbian Highflier – This is bred for endurance flying and originates from Belgrade, Serbia. It is a relatively recent breed and some evidence exists that the breed is descended from the *Illyrian Pigeon*. The crest is spade shaped. It's a small to medium compact pigeon.

South German Monk – It is always peak crested and clean legged or shell crested with medium length muffs blending with the hock feathers. The whole bird is pure white except for the wing coverts which are coloured.

Stralsunder Highflier – This emerged in Pomerania and is developed from the French *Cumulet*.

Strasser - This breed is also used for producing squabs as food.

Sverdlovsk – A blue grey mottle headed pigeon.

Szegediner Highflier (*Crested Tippler*) - Developed in Hungary.

Swallow - Originally from Germany, this breed has a shell crest and coloured cap. The colours are black, blue, silver, red and yellow.

 *Thuringen Field Pigeon* - This breed is known for several different colours and markings. Varieties include the *Thuringen Breast, Thuringen Monk, Thuringen Shield, Thuringen Whitetail* and the *Thuringen Spot*. The breed can be either plain or shell crested, but is always clean-legged.

Tippler – A breed of pigeon bred to participate in endurance competitions. It is thought to have originated in Congleton and Macclesfield in England, around 1845. There is a number of types of tipplers named after different breeders or the location they originated from. These include, *Hughes, Boden, Lovatt, Merredith, Shannon (Irish Delight)* and *Sheffield*.

Tumbler pigeon – These varieties have been selected for their ability to tumble or roll over backwards in flight. This ability has been known in domesticated breeds of pigeons for centuries.

Turbit – This breed is known for its peaked crest, short beak and frill of feathers on its breast. The body is white, with wing feathers in colours of black, red, blue, dun or yellow.

Ural Striped Maned – These birds were brought to the Ural region in the eighteenth century. They are small, white birds with a short neck. The mane may be shaped as an oval, a crescent or a triangle. The tail is red and has a white stripe 2–3 cm wide. The undertail is white or red.

Valencian Figurita - This breed is known for its small size and frill of neck feathers. It originated in Valencia, Spain.

Voorburg Shield Cropper - This breed was developed by C.S.T. Van Gink at Voorburg in Holland in 1935.

West of England Tumbler - This breed was developed in Bristol, England in the late 1800s and early 1900s.

Zitterhall (Trembling Neck, *Stargard Shaker*) - Originally developed in Pomerania. Zitterhals have curved swan-like necks that tremble or shake in a way similar to that of fantail pigeons.

3

Home Sweet Home – the Pigeon Loft

Pigeons need somewhere to live.

Houses for pigeons are generally referred to as lofts – so named because in the past most pigeons were kept in high buildings. There are as many different types of lofts as there are pigeon fanciers. Some people choose to build pigeon palaces that can accommodate hundreds of pigeons. Others construct more modest housing. Your loft needn't be huge or luxurious, but you need to bear in mind that not only will it be your pigeons' home, it will also be somewhere that you'll spend a lot of time.

The best loft is one that your pigeons and you are comfortable in. Build your doors wide enough for you to carry feed or crates without constantly bruising your elbows on the door frame. Similarly, the loft should be high enough for you to comfortably stand and move about in without nooks or crannies for your pigeons to hide or nest in.

It should contain openings to allow your pigeons freedom to exercise while allowing them to re-enter the

33

house without your help. Also, the loft needs to be constructed to keep the pigeons safe from predators and bad weather and to give them nesting places. Make sure you cover any openings with chicken wire, or with wire cloth if it's somewhere where your birds can get to (chicken wire can damage a bird's wings). Never leave the grain bin open because rats will get in and contaminate it.

And remember to fix a strong padlock on the loft door. There's no point in making things easy for burglars.

Location

Your loft can be placed almost anywhere as long as it's somewhere you can visit easily and frequently. Try to avoid building under or near trees, power lines or any close-by overhanging structures. These will give your birds a place to sit when they should be exercising or returning from a race. Ideally your loft would be somewhere that can be seen at a reasonable distance for 360 degrees so that your birds do not waste valuable race time trying to find their home. It's best to build your loft in an open area if possible.

Pigeons are happy to live in bustling cities, so you needn't worry about placing your loft somewhere quiet from their point of view. However, although pigeons are silent in flight and don't make a lot of noise, it's important to remember that annoyance by noise is highly subjective. Therefore, you will need to consider your neighbours when building your loft. Are they going to be OK with regularly seeing pigeons outside their houses? If you have a family, will they be happy to occasionally feed and let your pigeons out if you are not there? Your life will be so much easier if you consult your neighbours and, ideally, have their support.

You also need to think about the placement of the loft in relation to the birds' drop in. It's important to have as few obstacles as possible in the approach to avoid losing valuable time in a race because of the birds having to circle several times before making their landing. Ideally, your loft should be as far from other buildings as possible. Pigeons need to be able to see their surroundings. If possible, also avoid placing your loft near telephone or electrical wires. These are a common cause of broken legs and wings.

The colour scheme of your loft should be in keeping with your neighbourhood. In previous years, pigeon fanciers have painted their lofts in a whole range of fluorescent colours with the understanding that the brighter the loft, the easier their birds will be able to find it. This really isn't true though. My experience is that they will find the loft easily so long as they are trained properly and know the surrounding area.

Types of loft

You need to build your loft to suit your needs as well as your local climate. If space is at a premium, you will need to build up rather than out. If you have an existing loft, you can gradually make changes to get it just how you want it.

Purpose built lofts

Can be purchased from a specialist manufacturer and be found by looking in pigeon magazines or asking local fanciers. These can be bought with everything that you need to start but can be quite expensive.

Second hand lofts

Can be found through word of mouth or also in pigeon magazines. If you belong to a pigeon fanciers club, pass the word that you're looking for a loft. Lofts are also occasionally advertised on websites such as eBay, Craig's List and Gumtree. If you purchase one of these, be prepared to have to dismantle and rebuild it yourself.

Home build/conversion lofts

In my opinion this is the most rewarding and satisfying option for the new fancier. When converting an existing outbuilding or shed, it's doubly important to work out how many birds you will be able to comfortably house in advance.

Design of the loft

As previously mentioned, it does not take a lot to obtain a decent loft. There are a number of options available when you've decided you want a loft. You can buy relatively inexpensive loft kits, convert an existing shed or build one from scratch. Your pigeons really won't mind whether you spend £10 or £10,000 on their home, so long as it's comfortable for them.

In this book we shall be considering the back garden loft, as these are the most common and the easiest to set up.

The loft must:

- Be at least 20cm (8") off the ground
- Have a floor space of 4.3m x 1.8m (14' x 6'), to allow each bird 0.3m² (3' 3" squared) of floor space each.

- Contain enough nest boxes for all the birds when paired up.

- You should be able to comfortably stand up in each section of the loft and stand in the centre and touch all four walls without moving. This will ensure that you can spend time in there without it being awkward and that there is enough room for the pigeons without having to chase them when they need to be handled.

The style of loft matters doesn't matter, it's only important that the birds like it and you find it comfortable to use. You can use the following checklist to see whether a specific loft is suitable or whether you need to make any changes.

- Do the birds seem happy?
- Are they comfortable or do they fight all the time?
- Is there enough space?
- Is the loft easy to clean?
- Are there any nooks and crannies where the birds can hide?
- Can you move about the loft easily?
- Do you have enough storage space?

The biggest loft advantage is one where the fancier gets a good relationship with his birds by being frequently around, and in view, of the pigeons.

Maximising sunlight in the loft

You ideally want your loft to face east/south east to receive the morning sun, with windows on the east and west sides. If this isn't possible, you may be able to include an opening to allow all possible sunlight to enter in.

My preferred system for raising birds involves the darkening system (which is explained in *Best Systems* later in this book) and so the morning sun is very important. Sunlight will make your birds noticeably happier and more alert. I have found that using UV daylight bulbs inside the loft during normal sunlight hours, as well as giving the birds access to regular sunlight improves their mood.

Size

Pigeons from overcrowded lofts don't race to their true potential. Overcrowding can increase fighting and make the birds restless as well as making the air stale. It's very important to plan ahead how many birds you want to keep in total in the long run. You should allow at least 0.3m² floor space per bird and install enough nest boxes and perches for any new birds you are planning on having.

Over crowding will affect the health of your birds, make them unhappy and will affect their racing.

Floor

A wooden floor acts as a good insulator, stays warm and is smooth for scraping. (You can have one that can be unscrewed and replaced with wired floors during the off season.) The disadvantage of wooden floors is that they are harder to disinfect. Concrete floors aren't good for race lofts, because they are cold and retain moisture. However, they can be good for the breeding loft and you can concrete the race loft if you add heating. Wire floors cause problems as droppings beneath the wire can get damp and grow fungus. If you do have a wire floor, you need to be meticulous in regularly treating it for fungus and insects.

Roof

As you want your pigeons to get inside their loft immediately after landing, a flat roof is a bad idea. Ensure that you have a roof that the pigeons can't land on so that they are forced to land on the landing board.

Loft sections

You will need separate sections for your birds, regardless of what system you are using to train your birds. At the very minimum, if using the natural system (described in

Chapter 6: Breeding), you will need two sections to your loft: one for the mature birds and one for the new youngsters.

Remember that, if left to their own devices, pigeons will breed whenever they fancy it (instead of when their fancier fancies it!) so you will really need at least three sections to your loft: one for cocks, one for hens and one for young birds. It is always advisable to have a separate, smaller section as a bird hospital, where your sick or new pigeons can be quarantined until they are 100% disease free.

These partitions can be easily achieved by erecting a simple partition wall that splits your loft in half. It does not need to be particularly strong, just strong enough to support a door that you can move through and close behind you. To this end it can be made from a simple piece of plywood cut to fit the shed's dimensions with a hole for a door.

Aviary

An aviary is something that many pigeon fanciers do not feel a need for, which I find remarkable. It is a very easy structure to erect and gives the birds somewhere they can exercise or sun themselves when you're not around, and allows young or new pigeons to get used to their new homes.

It need not be complicated: just a wooden framework with a fine chicken wire around it is perfect. Making the top of the aviary from wood or shingles will help to prevent droppings from other birds – that could carry disease – from getting into the aviary. However, this needs to be balanced against the advantage of an open top aviary that allows rain and sunshine to enter all the parts of the aviary and for the birds to take rain and sun baths.

Damp air

The majority of this air will come in from coastal winds. The opening of the aviary must be facing away from the nearest coast or, if inland, away from the most prevalent damp winds, as shown on the next page. The loft entrance should face any direction marked with the broken line.

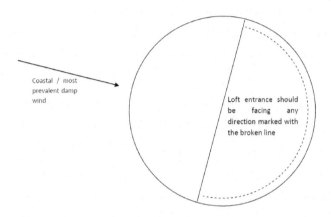

Coastal / most prevalent damp wind

Loft entrance should be facing any direction marked with the broken line

Ideal direction of loft entrance relative to coastal or most prevalent damp wind

Ventilation

This is one of the biggest mistakes new fanciers make: not having enough fresh air for their birds. The loft should never feel stuffy inside and should never be dusty and make you cough because of the particles in the air. Remember that pigeons don't mind the cold, but they do not like drafts.

Pigeons produce body heat and the sun beating down also produces heat. Your loft will need good ventilation with slow moving fresh air. Wind powered turbines installed into the walls work very well for this and are quite cheap.

Remember that warm air rises and cooler air is heavier than warm air. Your loft will need ventilation holes in the bottom to bring in fresh air and a vent in the roof for the warm, stale air to escape through (see the diagram on the next page). Having the loft raised from the ground will allow a steady flow of air beneath, but you will need to ensure that there is no way for vermin to enter the loft. Rats and mice will tunnel under your loft if given the opportunity, eating feed and potentially killing your pigeons.

Installing plastic louvre vents is the easiest way to help air circulation without allowing drafts to flow through. Cut a hole into the wall of the loft that is slightly smaller than the vent then screw it [the vent] over the hole... it's as easy as that! You may also want to consider putting hinged doors over the vents so that they can be closed over in the windier months if you find you're starting to feel a draft. Never have ventilation which will let rain beat into the loft.

The general rule is that so long as it doesn't smell musty in the loft then there is enough ventilation. Double check the quality of the air by asking a friend with asthma to stand inside for a moment – it's too easy for you to get used to conditions and not notice problems. Another test you can do is to put something smelly, like an air freshener in your loft. Come back ten minutes later and if you can still smell it, you might need to improve your ventilation.

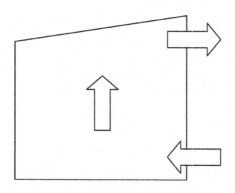

Example of circulation through a standard sloped roof loft. Air flow is created by air being warmed by pigeons and rising out of vents in roof.

Damp

Keeping your loft waterproofed should be a priority. Wet floors endanger your pigeons. Keep cleaning the loft using water for warm days or when the birds are out.

Humidity and temperature

Good airflow is critical, as is a dry environment because moisture in the air will make your birds ill. Lining the floor of your loft with sand with a lime base will remove a lot of moisture from the air.

Building your loft off of the ground as much as possible (at least 20cm) will help to prevent damp. Always ensure that the open space under the loft is covered to prevent your birds or predators getting underneath.

Ideally, the temperature in the loft must be above 10 and below 30 degrees Celsius, and the humidity kept below 65%. The humidity determines whether or not the birds rest at night. Rest is essential as without it your pigeons are susceptible to illness.

Droppings indicate humidity levels. A low humidity gives nutty brown droppings and a higher humidity will produce green watery droppings. There will be a rise in humidity at night because as the temperature drops the humidity rises.

You need to be especially careful about humidity if your loft is:

- Near open water
- Adjacent to large open low lying areas
- Without sunshine to dry the ground
- In a high rainfall area

Good insulation will help you to control temperatures and humidity inside your loft. Shutting the loft up at night or when it rains will also help, so long as that doesn't compromise air circulation.

Nest boxes

You need to allow one nest box per pair of birds. These are where your birds will stay when mated, sitting on eggs and rearing their young. Because pigeons are territorial about their nesting area, they get on better when each mated pair has two nest boxes of its own. Having fronts on your nest boxes will help to keep other birds out.

I've always found it best to have the nest boxes directly in front of the trap, where your birds will come in after a race. This means that the birds will be able to see their mates from the entrance and be more eager to enter to see their partner.

The size of the loft will dictate how many nest boxes you can comfortably fit. This will also dictate how many birds you can store at any one time as each pair of birds needs a box.

Pairs of birds = number of required nest boxes

(6 cocks and 6 hens would require 6 nest boxes)

Size of loft wall space = number of possible nest boxes

If you don't have the space for nest boxes then you don't have the space for more birds.

Building a nest box

The box should be about 60cm x 40cm x 40cm (23.5" x 16" x 16") with a landing section of around 30cm (12") (see the following diagram). It doesn't matter how these are made as long as they are at least as big as this and that they have a dowel rod front or some other way for them to be able to see out. They should, of course, have a *door by which the birds can enter*. It is important that the hen can be locked in the box when sitting on the eggs to avoid contact between her and the cock during race times.

You should fit as many boxes as possible into your loft to prepare for birds you will breed in the future. These extra boxes must be closed until a pair needs them because otherwise cocks that already have a nest will start to claim the empty ones as their territory.

The boxes should be lined with fresh sawdust treated with an anti fungal and insect repellent.

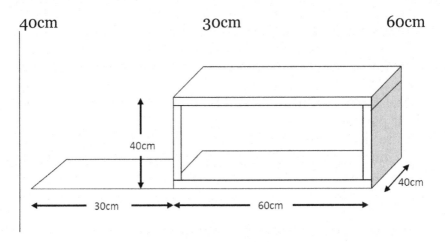

40cm   30cm   60cm

Dimensions of an example nest box

The nest bowl

You will need a nest bowl for each box, where your birds will lay and sit on their eggs. These can be purchased in a variety of materials but the pottery ones are best in my opinion. They are easy to lift up when you need to clean out the box, which should be done daily to avoid insects being attracted to any young bird droppings. They will have holes in the bottom to allow dust to escape and for ventilation to keep the contents dry.

Perches

Perches are particularly important for young birds. They will get their first flight experiences from flying up from the floor to a perch and will use this point to look around and get to know their surroundings.

V perches are easy to build, cheap, and simple to clean. However, it's easy for pigeons to soil the pigeons below them if you make the perches too short.

Box perches are more costly, but are excellent for controlling pigeons, because the birds can't jump from one perch to another when you want to catch them. It's also harder for them to soil the birds below. They shouldn't be too large – if they measure more than about 22 cm x 22 cm (9″ by 9″), the pigeons might get their own droppings on their perches.

Make sure that, like with nest boxes, you have plenty of perches for all the birds to easily have one each. It's best to have more perches than you think you'll need. Once a pigeon has selected its perch, it will fight to keep it. If you have more birds than perches, those without perches will feel left out and will have no desire to come home.

Landing board

Your landing board is in front of the traps and is the landing and taking off pad the birds use to enter and exit the loft. It should be big enough for the birds to land on and take off from comfortably. If it's too small, your pigeons won't be able to enter the loft easily, and they will be less willing to come when you call them. Ideally, it should be big enough for your youngsters to sit and look at their surroundings and become familiar with them.

Traps

A trap is basically a one way door for the pigeons to enter the loft and then be unable to exit, much like the basic principle behind a catflap. There are a lot of varieties available, each with their own pros and cons. It does not matter which trap you use, so long as your birds are quick to enter the loft when they return from a race.

A common form of trap is one that is bob wired where the birds push against the wires to gain access, but are restricted by the wires when trying to get back outside. Another form is called a Sputnik trap. This uses an opening set on an angle which is just wide enough a pigeon to enter by dropping through into the loft.

A very common way to lose vital seconds in a race is in the time in which it takes for your pigeons to re-enter the loft. If they are poorly trained or their trap is not designed properly then they may perch outside the loft instead of coming straight in. If a pigeon is loitering outside the loft then it may be that it has had a bad experience with being trapped previously. Make sure you do not cause the bird undue stress by handling it roughly or chasing it as this will make it reluctant to be handled in the future.

Before a race each pigeon has a rubber attached to its leg. At the end of the race this rubber must then be removed and clocked, requiring the fancier to catch each pigeon in order to do so.

Basic items your loft will need to be stocked with

Feed trough

It is very important that you don't leave food on the floor because it will spoil as well as attracting rats. The feeding section of this book will explain how best to feed. It is my opinion that it is better to feed birds at set times and not allow them free access to food. This will allow you to use food as an incentive for your birds to return after a race or training flight. Food troughs can be simply made from timber.

Water hopper

Birds need constant access to clean, fresh water which should be changed daily. I have always added half a teaspoon of liquid chlorine household bleach (5.25 percent sodium hypochlorite) to every gallon of water as this prevents bacteria from forming as easily. A cheap water hopper can be bought from any pet shop.

After a couple of weeks you will start to notice slime forming in the trough of the water hopper. This is easily removed with a slightly stronger mix of bleach and water and a sponge (remember to thoroughly rinse the hopper after each cleaning).

Keeping water outside in the aviary prevents it from getting spilled on the loft floors. During winter you can supply water inside the loft in a box with a light bulb to stop it from freezing.

Specialist grit

Pigeons need to have constant access to grit. Pigeons obviously do not have teeth so must digest food differently to us. They have gizzards that need small stones to grind the hard seeds to a pulp to digest the food. Commercial grit may contain minerals and salt that will be useful for your birds, but it's easy to make some yourself. The grit should be broken into pieces roughly pea sized; much smaller than that and their gizzards will break it down too quickly. Oyster shells can also be added to the grit as the calcium from the shells will help give them strong bones and firm eggs. Grit should be kept in a covered trough so the pigeons can't soil it.

Grain store

Somewhere grain can be stored and kept clean, dry and free from vermin is essential.

Sanitation and hygiene

Your birds need a clean environment to prevent illness and keep them happy: no-one likes to be surrounded by their droppings for long periods!

You want scraping to be as easy as possible. A flat, smooth floor and wide perches that can be brought out from the wall will make your task simpler.

Line the nest boxes with sawdust sprinkled with anti fungal/anti parasite powder, and clean droppings away a couple of times a week (maybe more often in the hot summer weeks).

Sand can look good on the floor, but it's a bad idea during the winter when pigeons may suddenly over engorge on it.

If you use straw for the pigeons to lie down on, make sure that it's always perfectly fresh, clean and dry. Black marks and a musty smell show that there's mould, which can damage the pigeons air sacs if they inhale it.

All feed troughs, water and grit containers need regular cleaning. Birds should be allowed constant access to fresh water, changed every day. The water hopper should be cleaned and disinfected at least once a week. The drinking vessel should be covered and not underneath any perches where droppings could fall into it. Young birds might try to bathe in the drinker, so make sure that the water outlet is not big enough for them to do so.

How often you clean your loft depends on how many birds you keep and how much space they have. If you have a lot of birds in a small area, you will need to clean more often than if you have a few birds in a large loft.

Ideally, the loft should be cleaned at least once daily. Regular cleaning helps you to keep a close eye on your pigeons. Any change in the droppings will be recognised early on and you'll be able to act on the problem quickly.

Keeping your loft clean is also important for your own health. Pigeon Fanciers' Lung (also known as Pigeon Breeders' Lung, Bird Fanciers Lung and Bird Breeders' Lung) affects numerous fanciers, but can be kept at bay with a few precautions.

The symptoms of this condition include flu-like symptoms of fever, chills, muscle ache, cough and or breathlessness four to eight hours after high level exposure to pigeons. No one knows precisely which pigeon materials cause the disease. Usually, the symptoms pass after about forty-eight hours. It can be very serious, so seek medical advice if you think that you've been affected.

Hygiene checklist

- Don't allow droppings to build up. The particles you can't see are the most dangerous and the best way to remove dust is with a vacuum cleaner. Every so often, wash the loft out with a hose while it's empty.
- Only use as much loft litter as is necessary.
- If you have respiratory problems, or allergies, wear a mask when in your loft.
- Wear protective clothing such as an overall or dust coat when in your loft and keep it outside the house expect when bringing it in to wash.
- Pigeon materials build up on the skin so remember to wash your hands with soap and water after handling birds. Train your pigeons to be used to you walking around the loft so that they don't panic and fly about when you enter.
- Provide baths - preferably outside – for your pigeons.

Sensible hygiene and regular cleaning are all that are needed to keep you and your birds in optimum health.

4

Feeding

Proper feeding and nutrition is essential for your pigeons to maintain their health. An average pigeon eats about twenty eight grammes of food per day – about twice as much if it's feeding its young. A full-grown pigeon will eat about 500 grammes of grain each week. Birds that exercise more need more food.

Pigeons are seed eaters and need a balanced diet. Eating just one kind of seed will cause them to become sick and weak. They tend to pick out what they like, and leave the

51

rest, and will not get the nutrition they need, if you let them pick out the seeds they like. Avoid offering foods normally reserved for people, such as bread. The complex carbohydrates in bread offer no nutritional value.

Ideally, a pigeon keeper should try to emulate the wild birds' diet as best as they can. Pigeons like grain better than anything else, and will east a wide variety of types. Some grains are better than others and others should be avoided. Dried field peas make a good staple food and pigeons really like them. Peas are high in protein and encourage strong muscles, bones and plumage. Corn is probably the pigeon's second favourite food. It makes a good winter food as pigeons find it easy to store in their bodies as fat. Because of this, you need to be careful not to give too much corn to your birds, especially if they have limited exercise. Wheat is common in pigeon mixes, and most pigeons prefer the darker red wheats to the soft white kinds of wheat.

Store your feed in a covered container that will keep mice or rats out and remove any wet or old food as soon as possible to avoid the build up of bacteria, moulds or toxins.

An alternative to a grain mix is pigeon pellets. Given the choice however, pigeons tend to prefer more natural grains.

The following is a list of my food mixes that I offer as an example. There is no definitive right and wrong food mix for birds as long as they get a variety of seeds to ensure a balanced diet. Along with these mixes it is a good idea is to add some chopped green leaves (lettuce, cabbage, spinach etc.) to their food as well, especially in the winter when they may not be able to get enough vegetation outdoors.

Two weeks before mating:

- 25% beans (peas, tares)
- 25% maize
- 25% wheat

- 25% mixture of mainly barley with buckwheat, rye, oats and linseed

Contains good levels of nitrogen and is rich in hydrates and carbon. It will also help bring your bird back to full health after the strains of racing. One meal per day is sufficient but a handful of linseed in the mornings will help lubricant the passage for egg laying.

During rearing

Fourteen days before rearing: Pigeons should be left to fast without food for an entire day.

Thirteen days before rearing: Given a good feeding of only linseed in the afternoon.

Twelve days before rearing: composition of feed should be changed gradually until arriving at the composition:

- 45% beans (peas, tares)
- 35% maize
- 15% wheat
- 4% buckwheat and linseed
- 1% of husked barley, rice, millet, canary seed and (a small amount of) bread given as a dessert.

At this stage they will require two meals per day.

Racing season

As soon as the youngsters are weaned, the pigeons will be ready to go into training.

- 30% tick beans
- 15% tares
- 5% peas
- 40% maize
- 5% wheat

- 5% small, mixed seeds*

* (buckwheat, linseed, millet, rice, colza, turnip, hemp, barley and husked oats)

In July/August give the pigeons you intend to race a very small amount of hemp seed.

Three meals a day is advisable to get them in top form - ideally at 6 am, 12 noon and 5 pm. (If this is not possible, then two meals at 8 am and 4 pm.)

Moulting (after their second round of eggs)

- 35% tick beans (peas, tares)
- 40% maize
- 20% wheat
- 4% linseed
- 1% barley

Two meals a day.

NOTE: The proportions given are only an indication: don't worry about following them to the letter. This is a great starting point, but you should test and change the amounts in small increments to see what works best for you and the breed of your pigeons. Some birds needs stimulants like hemp seed while others do not need them.

Note on winter

Do not allow your birds to mate between September and February because they will need to conserve their strength over these cold months and prepare themselves for the racing in spring.

It is advisable to start increasing the amount of barley in their diets during winter. It is a natural blood purifier and will help ward off disease and keep your birds healthy.

Other concerns

AVOID SUDDEN CHANGES

Take two or three weeks to change from one plan to another. A sudden change in the feeding plan will cause your birds to become ill. This also goes for changing to different types of grain.

Buy your stock of feed for the new plan three weeks in advance, gradually incorporate it into your current feeding plan and feed simultaneously from old and new stocks. This is especially important during racing season where you cannot afford for your birds to become ill from a sudden change of diet.

YOUR BIRDS NEED GOOD GRAIN

Whether you're purchasing ready made mixes for your birds or making your own mixes up you need to ensure that they are good quality.

Avoid purchasing grain gathered under bad conditions at all costs. If possible, ask other local fanciers where they buy their grain supplies from and go and check how it's stored there.

Grain should be kept in a dry, well ventilated place with proper precautions against rodent infestation, cats and wild birds. Any droppings or diseases from rats or mice will cause your birds easily avoidable illness.

Ideally, grain should be stored in a bin with a metal sieve at the bottom to allow ventilation and be sifted occasionally to avoid heating.

Do not worry about the age of the grain you give your birds. Some people report that newly gathered grain gives their birds diarrhoea, but I believe this is more to do with the new grain not being gradually introduced into the food mix.

How do I know when my pigeons have eaten enough?

To ensure the health of your birds it is best to never give more food than they can eat. Change their water daily and do not over feed. Not following these two rules will eventually cause your birds to become ill.

Although regular feeding by the fancier themselves is the ideal (if only because it solidifies the bond with the pigeons), if time is not available then a seed hopper is a good purchase to keep the feed clean.

NOTE: Never feed your pigeons directly off the floor, always use troughs or something else that can be emptied or cleaned after use. Food left out attracts rats which will spread disease.

DO NOT OVERFEED

Every morning handle your birds and feel their crops. If there is any food in them then it is a sign that you fed them too much the night before and you should cut their food rations down slightly. Similarly, if your pigeons do not rush to the sound of you rattling their food tin, or calling, then they are also being fed too much. If you feed your birds and noticed there is still feed left after ten minutes, you are feeding too much.

Pigeons that receive too much food will often fall sick and, in fact, birds that are slightly underfed for periods actually seem to perform better which is why I would recommend giving them a restricted diet (as long as it is one that is healthy and balanced with a variety of grains and seeds).

You might worry that you are feeding your birds too little initially but there is an easy way to check: give them a bath. If they really want to get in the bath then they're being fed enough, if they seem lazy and bored then they don't feel well.

Supplements

Probiotic powder or capsules added to drinking water helps to replace good gut bacteria and keep your pigeons healthy. Apple cider vinegar can also be added to their water once a week to help to prevent diseases like coccidiosis, canker and crop candida. Garlic is useful as it acts as an antibiotic and keeps the feathers looking nice and keeps internal parasites away. A little cod liver oil mixed into the feed acts as a good tonic and small quantities of linseed meal can give added gloss and lustre to the birds' plumage.

Water

Pigeons need to drink after they eat to soften the food for digestion and to regulate their body temperature. Unlike other birds, pigeons suck up water. Water is even more important to pigeons when they are feeding their young. When the weather is hotter, they will need more water to drink. Drinking water should be changed every day if possible, and supplied in such a way that the birds can't foul it with their own droppings. Pigeons need a water container with at least 3.5 cm of free standing water that they can drink from.

Grit

As pigeons don't have teeth they need to eat sand, stones, shell, salt or grit to break down and process food. Their food is swallowed whole and stored in a sac at the base of the throat called the crop. The food then passes through the digestive tract to the gizzard where it's broken down. For this to happen, your birds need a supply of grit. Hens that are getting ready to lay especially need grit to increase their calcium levels to create stronger egg shells. Pigeons that have free access to the outside world will find their own grit.

In a nutshell

Follow the feeding steps I set out and perform the following checks to see how your birds are finding their diet:

- If they start to seem very interested in eating their own droppings then they either are not being fed enough or may be short of vitamin B12.

- If they don't want to bath then they don't feel well which may be due to not enough food (or another illness).

- If normally compliant birds are hard to pick up after feeding time then they have excess energy and are being fed too much.

TIP: Feed very sparingly at the beginning of the week before a race and gradually increase the amounts until the end of the week, with them being fed the full amount on race day. I would not recommend this if you are racing the same birds week after week because they will need food to build their strength up after racing. Used every other week this can be quite effective.

Feeding

How to feed

Throw your birds a handful of feed into their feeder (never on the floor) at a time, not all at once, and throw it down just in front of you. This will force the pigeons to come close to eat and get used to being in close proximity to you. Those birds that are not quite tamed will initially hang back and their fear of you will result in them not eating. Do not be alarmed, hunger will soon tame them and within two days all your birds will be eating out of your hand!

Do NOT feed your pigeons late in the evening (i.e. after 5 pm) because this will make them restless and unable to settle down for the night and get enough sleep.

Golden feeding rules

1. Watch your pigeons as they eat.

Feed your pigeons a handful at a time. When two or three pigeons stop feeding and go to drink then don't throw the remaining birds any more food. Those birds that are still left will be able to eat any remaining seed in the trough and, if still hungry, will be more eager to eat next feeding time.

2. Do not feed your birds before they go out.

Let your birds out whatever the weather (except for in snow and fog) and while they're gone take the opportunity to clean their loft and give them fresh water. Recall them after half an hour or so and feed them when they are back. Those who do not return when you first call get no food and quickly learn to return straight away.

The calling system is very straight forward and crucial to well tamed pigeons: decide on what your food call will be. This is what you will do before each time you feed your birds. It can be a rattle of a tin or something you call out, it doesn't matter, so long as it always stays the same. Your birds will quickly come to associate this with food and will hasten into the loft the second you call.

3. Don't change your system while the pigeons are bringing up their young.

Giving the birds extra food in the nest will not help the youngsters grow up faster; it will just teach them that they will get food regardless. Much like with young children, you need to have a set of rules that you do not deviate from in order to get the most out of them.

***

Feeding time is one of the most important times of the day. It allows you to get to know your pigeons and gain their trust.

5

Good Health

There are numerous disease hazards that your pigeons can be exposed to. Many of them are beyond your control and often you may not realise that your bird has been exposed. Your pigeons may mix with feral birds or other flocks and, while they are out, what they eat or drink and where they roost is beyond your control.

By following three aspects of health management, you can deal with problems efficiently.

- Prevention - This includes vaccination against conditions, exposing the birds to possible pathogens to increase their immunity and practising good hygiene.
- Monitoring - The most obvious first sign is poor performance. Overt symptoms may not emerge until the disease has taken a firm hold. Keep an eye on food and water consumption, plumage, weight and body condition, exercise performance, reproductive efficiency and droppings.
- Diagnosis and treatment.

It's helpful to keep a stock of medication in your loft against common conditions in case of emergency. Be especially careful when adding birds to an established flock since they may be carriers (quarantine them for thirty days before adding them).

Because pigeons are highly stressed and susceptible to infection during shipping and racing, many people choose to use antibiotics for a few days each week to prevent bacterial diseases.

Common signs of illness:

- Sits apart from the flock looking hunched up and depressed
- Fluffs feathers up
- Has difficulty keeping eyes open
- Stays on the ground when the flock flies off or at dusk.
- Walks in circles.
- Throw seeds in the air.
- Totter backwards
- Has fits
- Uses its wings to walk with (because its feet have become tied together)
- Unable to fly

Handling a sick pigeon will not put your health at risk, although it's advisable to take sensible precautions like washing your hands after handling it.

Do's and don'ts

- Do keep sick pigeons isolated
- Don't place caged or baby birds in direct sunlight or in front of a hot stove
- Don't squirt water into a bird's mouth. They can't cough and can die if it gets in the windpipe.
- Don't feed a pigeon as soon as you get it. Rehydrate it first in case its digestive system isn't functioning due to starvation or sickness.
- Don't hold a pigeon too tightly.
- If there are cheesy growths in the mouth do not remove or dislodge them. It could start a fatal bleed or suffocate the pigeon.
- If there is an egg lodged in the cloaca, don't break it or try to remove it.
- Don't release a pigeon just because seems to have recovered. It needs to be well enough to find food, shelter and avoid predators. Err on the side of caution.

In an emergency:

Three basic steps to follow when a pigeon is sick or injured:

Heat – Warm it gradually to a normal body temperature. Don't give a cold bird fluid or food. Unless there is a critical situation, for example heavy bleeding, the bird should be covered and placed on a heat source for at least 20-30 minutes. Heat sources include a towel lined heating pad, set on low, a towel lined hot water bottle or a low wattage lamp, directing the light into the cage. In an emergency you can use an old sock filled with rice and warmed in a microwave.

Isolation – This will allow the birds condition to stabilise as well as protect your other birds in cases of infection.

Hydration - Fluids should be given after the bird has been warmed and examined. All fluids should be warm or at room temperature. Symptoms of dehydration include lethargy, dry, flaky skin, dull eyes, non-formed droppings, and a sticky membrane in the mouth. If the bird is alert, it can be rehydrated using an eye dropper and putting drops along its beak every few minutes. A rehydrating solution should be used. A rehydration solution can be made up from half a litre of water, half a teaspoonful of salt and half a tablespoon of sugar or glucose. Only use plain water if nothing else is available. If the bird is not swallowing or fully alert, it needs to be given fluids under the skin by a vet.

If a pigeon is seriously ill, you might have to feed it by hand. A seriously ill pigeon can be fed by tube to ensure that it has the vital nutrients that it would not take in on its own. Commercially made products come as a powder, to be mixed with water. A good mixture can be made by grinding pigeon pellets and water together in a blender.

You will need a feeding syringe, which has a large opening at the end to allow the passage of thick feeding solutions. Plastic syringes are best as they are durable and easily cleaned. Feeding tubes are either made of stainless steel or rubber. Rubber tubes will move with the pigeon if it struggles, preventing damage, and so are much safer to use. It's easier if you have an assistant hold the pigeon so that both of your hands are free. Be careful not to put any pressure over the crop area as this can cause vomiting. Lubricate the feeding tube with Vaseline before use to help its passage into the esophagus.

The pigeon's neck is straightened out vertically while the beak is opened. Once the mouth is open, it's easy to see the opening into the windpipe on the lower portion of the mouth just behind the tongue. The tube should be placed past this opening to the rear of the mouth on the bird's right side. Don't force it if you feel resistance. Slowly inject the solution into the crop. If reflux occurs, the pigeon should be released immediately to let it clear its throat.

Healing a broken wing

A bird with a broken wing will usually be grounded and hold one wing lower than the other.

Isolate the bird - A pigeon that is otherwise healthy will try to run away from you. It needs to be confined in a small place.

Examine it - A broken wing will hang differently, lower or at an awkward angle. It may drag on the ground. Wash any open wounds and apply an antibiotic.

Immobilise the broken wing - Gather together gauze that sticks to itself and some scissors. Cut about 15 cm of tape. Secure the broken wing against the body in its natural position with your hands. Wrap the tape on the outside of the broken wing and around the body under the healthy wing. Don't wrap it too tightly. Make sure that its feet are behind the tape (on the tail side) and that they do not get caught in the tape. This can be easier if someone else holds the bird's feet back and good wing up. Secure the tape to itself.

Let the pigeon free to move around. It might fall over and be awkward for the first day. .Keep the bird in a safe place and confined to a small area. Check it frequently to make sure the tape is still secure. You will need to leave the tape on for two to four weeks. If it gets soiled, change it about once a week.

Common diseases

Adeno virus

This virus is always present in a young pigeon and only erupts in birds whose immune systems fail. Type 1 affects young pigeons and causes vomiting and diarrhoea. The chances of recovery are high, although it's often accompanied by E coli, which can cause additional problems. It can sometimes be successfully treated with antibiotics. Type 2 is contracted by older pigeons and strikes the liver. Most affected birds die within twenty four hours. Some birds have fluid yellow diarrhoea and vomiting before death. There is currently no vaccine.

Canker (trichomoniasis)

Canker is a parasitic disease that mainly affects the throat and is caused by the single celled parasite trichomonas gallinae. Birds become infected through their drinking water and from regurgitated food or crop milk. Canker leads to yellow spots on the throat where the parasites have altered tissue into a cheese like substance. Symptoms can be more subtle and may manifest by slight inflammation of the throat and strings of mucous. Affected pigeons in a loft may cease to feed, become listless and ruffled in appearance, and lose weight. They can have difficulty closing their mouths because of lesions and may drool and make repeated swallowing movements. Watery eyes, diarrhoea, increased water intake and respiratory distress are also possible. Sometimes, there are no symptoms at all. However, even mild cases can damage a pigeon's performance and lead the way to more serious problems. Canker is diagnosed by a throat swab.

Be careful when treating canker during the breeding season as the drugs can have an adverse effect on fertility in the cock. It's best to avoid those drugs after pairing, before the eggs are laid, and when the breeders are feeding young. The best time to treat is before hatching. Canker is very common and it's a good idea to medicate all birds that share a sick bird's water and feed to be on the safe side. Your birds should be given a rest from medication during the off season to decrease the risk of resistance to the drugs and to allow them to develop a level of natural immunity.

Circo virus

Often referred to as pigeon AIDS, circo virus damages the lymphocytes in the blood, which are closely associated with the immune system. This leaves the pigeon susceptible to secondary infections. Usually, circo virus kills very young pigeons and strikes older youngsters that have already moulted three or four flights. Pigeons with the virus have a yellowish discharge dried on the beak, are reluctant to move, thin, dehydrated, have difficulty breathing and have no appetite. It can be hard to diagnose as the symptoms can be caused by secondary infections. If you have the virus in your loft, isolate any sick birds immediately while treating. Probiotics help birds to resist the disease.

Chlamydia

This microorganism is found in many pigeons. There are numerous strains that vary widely in their capacity to cause disease. Lofts often have resident strains to which the resident birds have developed immunity. When they become stressed the chlamydia flares up. If chlamydia is in an egg, the developing embryo is weakened and can die.

The tonsils may be inflamed, a thick white mucus may extend into the throat from the windpipe or from the slit in the roof of the mouth, the top of the windpipe may be red and inflamed, the beak at the nostril opening may be wet, the cere may be slightly discoloured and the gums or the muscles may be bluish. Chronically infected birds show delayed recovery after a race and develop green droppings because of damage to the liver. Medication can be given before mating to decrease the level of chlamydia in the stock birds' system.

Coccidiosis

This intestinal disease is very common. It is in most pigeon lofts, but only a very serious infection causes visible clinical signs. Coccidia is a small protozoan that lives in the wall of the bowel and its eggs are passed in droppings. As most pigeons build up resistance against this parasite, treatment is only necessary when there is a serious infection. Symptoms include disturbed digestion, flat and watery droppings, weakness, delayed growth and pigeons that are pale and droopy and tend to huddle. Coccidiosis is rare in birds under three weeks or in mature birds.

Escherichia coli (E. coli, collibacillosis)

This bacterium commonly infects pigeons. It is normal in pigeon faeces, but some strains can be severely pathogenic. As the bacteria can manifest in any part of a pigeon's body, symptoms of an infection can mimic symptoms seen with other diseases - weight loss, diarrhoea, dead in shell, joint abscesses and head tilt. E coli can affect young and old birds, and is usually associated with stress such as racing, breeding and overcrowding. There are no commercially available vaccines, but antibiotics are effective.

Haemoproteus

This blood parasite is transmitted from bird to bird by pigeon flies and causes anaemia.

Mycoplasma

Mycoplasma don't directly cause disease although the organisms do superficial injury to the lining of the respiratory system. This enables secondary organisms to become established. The symptoms depend on the part of the respiratory system affected. Mycoplasma tends to cause inflammatory changes in the top 20-30% of the windpipe, causing mucus to accumulate. When the air sacs are affected, the bird can't breathe properly and even moderate exercise is tiring. The gums and muscles can turn blue and the birds need to drink more. Mycoplasma is more likely to cause disease when the birds are stressed.

One eye colds

Eye colds are usually associated with a physical injury affecting the eye. They can also be caused by improper ventilation, drafts or dampness in the loft. Symptoms include a watery or mucous discharge in the eye.

Parasites

Lice and mites can infest your pigeons. Lice feed off feather debris and skin flakes causing the birds to become restless. Mites are a more serious problem as they feed off blood. Some species are found on the birds and others in cracks and crevices of the loft. Symptoms include birds becoming visibly pale and weak. There are many good and safe insecticides available to combat them.

PMV 1 (Paramyxovirus 1)

This deadly disease is transmitted by direct contact. It causes inflammation of the kidneys and infected pigeons produce a massive amount of urine and the white portion droppings is replaced by a pool of water. The pigeons will drink a lot to keep up with urine production and the loft will get wet. Also, the virus causes neurological signs such as lameness, dropped wings, twisted necks and inability to fly. Vaccines are available, but they must be used before the birds are exposed to the virus.

Paratyphoid

This is the name given to the disease caused by salmonella. Birds under six weeks of age develop severe gastroenteritis. The pigeon's eyes can become glassy and, if the disease progresses, it leads to emaciation and greenish diarrhoea. It can also affect joints leading to hot swollen joints in the wings or legs, or inflame the membranes round the brain leading to loss of balance and head tilt. The bacteria can cause sterility in both sexes and can contaminate the egg prior to laying. It is one of the two principal infectious causes of dead-in-shell and nestling death. Many birds exposed to the bacteria show only mild or no signs and develop into asymptomatic carriers. Infected pigeons need to be isolated and treated with an antibiotic.

Pigeon pox

This is caused by a virus carried by biting insects. It enters the bloodstream of the bird, and within five to seven days, small whitish wart like lesions appear on the head, feet legs and beak areas. The birds develop a fever before lesions appear. The lesions are firmly attached to the skin and hard to peel off although in time they dry and fall off. Vaccines are available, but there are no remedies.

Respiratory diseases

These are very common in pigeons and are the major cause of poor performance and pigeon loss during the race season. More than three sneezes within five minutes is a significant indicator of early respiratory disease. Young birds under stress are most at risk. Several organisms can be involved. The respiratory system can be infected by chlamydia, mycoplasma, bacteria, fungi, viruses and mites. Stress is always a big factor. Treatment includes reducing stress and any environmental triggers.

Worms

These internal parasites weaken the birds and can lead to a delay in laying, reduced growth rate, delayed weaning and poor food conversion. Roundworms and hairworms live as parasites in the pigeons' intestine. They extract important nutritive substances from the digested food. Young pigeons are the most susceptible to them.

Good Health

Poison

If birds are allowed to freely roam the garden then make sure they do not spend too much time on the grass. Pigeons will become sick if they are exposed to certain garden substances. It is really important for pigeons to not eat their droppings, which pigeons seem to love doing even though it makes them very sick very quickly. Some common substances can be harmful – if not fatal – to your pigeons. Never feed food meant for humans to your birds and be particularly careful with the following.

Chocolate affects a bird's digestive system, causing vomiting and diarrhea. In the long term, it affects the bird's central nervous system, causing seizures and eventually death. Caffeine is associated with increased heartbeat, arrhythmia, hyperactivity, and cardiac arrest in birds.

Apple seeds contain trace amounts of cyanide. The fruit of the apple is fine.

Avocados can cause cardiac distress and eventual heart failure in some birds. Although the toxicity of avocados is under debate, it's best to keep avocado products as far away from your birds as possible.

Onions can cause vomiting, diarrhea and other digestive problems. Prolonged exposure can lead to a blood condition called hemolytic anemia, followed by respiratory distress and eventual death.

Alcohol depresses their organ systems and can be fatal.

Mushrooms have been known to cause digestive upset in birds. The caps and stems of some varieties can induce liver failure.

Tomato stems, vines and leaves are highly toxic to birds.

Salt is fine in small amounts, but too much can lead to a host of health problems in birds, including excessive thirst, dehydration, kidney dysfunction, and death.

Raw dry beans contain a poison called hemaglutin which is very toxic to birds.

The following plants are all toxic to birds and are best avoided.

Amaryllis
Andromeda japonica
Apple seeds
Apricot pit
Asparagus fern
Autumn crocus
Avocado (fruit and pit)
Azalea
Baby's breath
Bird of paradise
Bittersweet
Branching ivy
Buckeye
Buddhist pine
Caladium
Calla lily
Castor bean
Ceriman
Cherry (wilting leaves and seeds)
China doll
Chinese evergreen
Christmas cactus
Christmas rose
Chrysanthemum
Cineraria
Clematis
Cordatum
Corn Plant (all dracaena species)
Crown vetch
Cyclamen
Daffodil
Daisy
Day lily
Devil's ivy
Dieffenbachia

Dracaena palm
Dragon tree
Elephant ears
Emerald feather
English ivy
Fiddle-leaf fig
Flamingo plant
Foxglove
Fruitsalad plant
Geranium
German ivy
Glacier ivy
Gladiola
Glory lily
Hawaiian ti
Heavenly bamboo
Hibiscus
Holly
Hurricane plant

Hyacinth
Hydrangea
Impatiens
Indian laurel
Indian rubber plant
Iris
Japanese yew
Jerusalem cherry
Kalanchoe

Lilium species
Lily of the valley
Marble queen
Marijuana
Mexican breadfruit
Miniature croton
Mistletoe

Morning glory
Mother in law's tongue
Narcissus
Needlepoint ivy
Nephthytis
Nightshade (solanum species)
Norfolk pine
Oleander
Onion
Peace lily
Peach (wilting leaves and pit)
Pencil cactus

Philodendron (all varieties)
Plum (wilting leaves and pit)
Plumosa fern

Poinsettia fern

Pothos (all varieties)
Precatory bean
Primula
Privet
Rhododendron
Ribbon plant
Sago palm
Schefflera
Sweet pea
String of pearls/beads
Taro wine
Tomato plant (green fruit, stem, leaves)
Tulip
Weeping fig
Yesterday, today, tomorrow plant
Yucca

Seeing your vet

The only way to know if some conditions exist in your pigeons is to have tests - faecal, throat swabs and blood tests - done by a vet. This way you will:

- Know if you have a problem.

- Not treat needlessly

- Know the proper drug to use

- Know the proper dosage of the drug

- Know how long to treat

It could actually work out much cheaper in the long run than guesswork.

6

Breeding

Natural selection, although it has worked for millions of years, is not an advisable path to take when breeding your pigeons:

Leaving your pigeons to their own devices would result in the survival of the fittest with the strongest cocks winning the best hens. This sounds like a good idea but the resulting injuries from fighting and the unrest within the aviary would upset your racers, who need equilibrium and routine to flourish. Conversely, the 'strongest' cocks may not be the best breeders and so their chosen mating would undermine the overall quality of your loft.

A harmonious loft is a successful loft. A disharmonious loft is a waste of space. Allowing your pigeons to mate freely would result in your best long distance racers mating with your short distance speed racers, giving you massive confusion when it came to the races and eggs would be hatched whilst other birds were still finding a mate.

Your main goal in breeding should always be to completely remove all genetic defects (minor and major alike) and to increase the amount of desirable traits to give you killer racers.

Advantages to natural selection

- It is a lot less time consuming
- It takes no effort on the fancier's part
- As the birds are allowed to make their own choices there will be less fighting between cock and hen.
- The birds will instinctively go to nest

Disadvantages to natural selection

- Any good bird produced will be a fluke
- The overall quality of your birds will go down because there's no guarantee that two birds mating will not have the same defect, resulting in an offspring with that problem amplified.

In my opinion, it is far better and more productive to choose the mating pairs than it is to let them choose for themselves.

Your main aim for breeding should be to match your finest birds together that have none of the same defects to produce a bird with all of their strengths with the weaknesses removed. As you will see it is a process that even well seasoned fanciers get wrong and something every fancier should understand inside out if they want the best bred birds.

The problem with mating your two best racers together is that there will be a period during the racing season when they will both be feeding offspring. If, during a race, one of them is killed or does not return then you have the problem of the partner bird (who is also one of your best racers) no longer having a mate. This leaves you with the dilemma of racing your bird unmated, resting her till the next season or the time consuming process of finding her a new mate.

In my opinion it is better to mate an old experienced hen who is being specifically trained for one five hundred mile race with a young cock (on the widowhood system, explained in Chapter 17) that you believe to be a good bird. As always, it is crucial that at least one of your birds comes from a good line as mentioned in the section on stock birds.

Regardless of the system that you put your birds on, it is VITAL that you keep accurate records of all breeding, as you will see in the section on inbreeding. A lot of fanciers seem almost afraid of paperwork and this is one of the easiest things you can implement that will give you the information to get the best results.

What is inbreeding?

Inbreeding is the process of mating your pigeons with other pigeons from the same family (i.e. brothers/fathers with sisters/mothers).

What is crossing?

The act of mating two pigeons that have no common ancestry.

What does inbreeding do?

Imagine you've got a trait in your birds that you want to replicate (which you obviously will have). If both the mating birds are from a common ancestor, e.g. the same parents, then basically, inbreeding increases the chance that this trait will be passed on. However, the same goes for negative traits: two related birds with the same bad quality will more likely reproduce an amplified version of this in their offspring, if mated.

The mechanics of breeding

The characteristics of living things, in this case our pigeons, are controlled by their genes which are handed down from their parents. It is not important to know exactly how these work, only that there are always two genes for every characteristic; one inherited from the mother and one inherited from the father. These two genes may be the same or they may be different, i.e. a pigeon might have a gene for having all red feathers (from the father) AND all white feathers (from the mother). Does this mean your youngster will have pink feathers because the genes mix together? No. Why?

Let me explain with an example:

- My mother has blonde hair, which means she has a gene making it that colour.
- My father has black hair, meaning he has a gene makes it that colour.

Would I have been born with a mixture of the two and have gray hair? No, I have black hair.

There are two types of alleles that make up our genes:

dominant and **recessive**. If you have a dominant allele
and a recessive allele for the same gene (like in my case
hair colour) it will always be the dominant allele that is the
decider. The recessive allele will stay dormant and may
surface in a future child.

Important definitions

Moving on from this, there are four words that you need
to understand (that sound difficult, but are relatively
straightforward):

Gene/genotype - the genetic make-up of an organism
with regards to a particular trait, such as wing length,
colouring etc. Each gene is comprised of two (or
sometimes more) *alleles*.

Alleles - a pair of different versions of the same gene
received from each parent. In my previous example I
received the allele for blonde hair from my mother and
the allele for black hair from my father. The dominant
black hair allele caused me to have the gene for black
hair and so I have the black hair *phenotype*.

Phenotype - the physical appearance of the gene. My
phenotype would be 'black hair' for this gene. If,
hypothetically, all my alleles for every gene were
identical (so I would have two black hair alleles, two
brown eye alleles etc.) then my phenotype would match
my genotype because my physical qualities would match
my genetic qualities with no hidden recessive genes. In
this case I would be completely *homozygous*.

Homozygous - if any two pairs of alleles in one of your
birds are exactly the same then they are homozygous for
that gene. This means when they breed then will
definitely pass the gene onto their offspring.

Heterozygous - if any two pairs of genes are different then there will be a roughly, 50% chance that this gene will be passed on to the offspring.

Why do we want to increase homozygosity?

Let's say you have a cock that is a strong long distance racer but we want him to fly faster. It would make sense to imagine that by pairing him with a hen that has won various shorter distance races that their offspring would be a faster long distance bird. However, if these two birds are completely heterozygous then there is little chance that the offspring will inherit a mixture of these abilities.

However, if (for the sake of example) we imagine there is one gene for being a great long distance racer and one for being a fast racer and the cock and hen are homozygous for them both respectively then by mating them we will definitely see the two attributes coming together. This is because there would be a 100% chance that the long distance gene would be passed down and a 100% chance that the speed gene would be passed down.

Of course, real life is not as straightforward as this and there will be many, many different genetic factors into pigeon performance, but this is the general idea behind striving for uniformity in your loft.

Brief explanation of any form of breeding

Let us examine the following table. A cock and hen both with a dominant gene B and a recessive gene b. Let us, for sake of argument say that B is the gene for a red head and b is the recessive gene for a blue head. Both these birds will therefore have red heads, with the B gene dominating the b gene and we want to make more birds like them.

Breeding

	Cock		Hen	
	Bb		Bb	
Gametes	B	B	B	B
Zygotes (offspring)	BB	Bb	Bb	bb

Showing the breeding of two birds, each with dominant gene, R, and recessive gene, b.

NOTE: A gamete is a single sex cell, either a sperm or an egg which carries one of the genes but not both. A zygote is a developing embryo that will eventually hatch out of an egg.

These arrangements are an average, i.e. there is a 25% percent chance of a **BB**, 50% chance of **Bb** and 25% chance of **bb**.

There are four possible combinations of the gametes:

BB - a homozygous red-headed bird that will always pass on the red-headed gene to its offspring

Bb - a heterozygous red-headed bird with both the dominant red-head gene and the recessive blue headed gene. This bird will pass his head colour on 50% of the time, like his parents.

bb - a homozygous blue-headed bird that will always pass on the blue-headed gene to its offspring.

The **bb** is the only bird that we would know is homozygous because in order for a recessive gene to manifest itself it must not have any competition from a dominant gene.

This bird will no longer be used for breeding as we want to phase out the blue heads.

Out of the three remaining we would see three redheads and know that one of them must be a homozygous red, the **BB** (it wouldn't be as obvious as it is in the chart, we would not be able to distinguish between **Bb** and **BB** because we only have appearance and performance to go on without breeding further).

To increase the amount of homozygous traits that we want, we must move on to inbreeding.

Inbreeding – how does it increase the overall standard of the loft and how much is too much?

By mating the three red-headed youngsters back to their parents and watching the inbred offspring produced we can see which birds are homozygous and which birds aren't.

For example: if from our initial pairings we select one of our offspring and it is a heterozygous **Bb** then from the offspring we would again see a blue headed **bb** produced. However, if the selected bird is the homozygous **BB** then there would never be any blue headed youngsters and instead 50% of the new offspring would be homozygous.

Let me explain with another table:

	Cock		Hen	
	BB		Bb	
Gametes	B	B	B	b
Zygotes (offspring)	BB	BB	Bb	Bb

Breeding

All the new offspring would appear red due to the dominant **B** gene. By back breeding all these birds to their heterozygous mother we can deduce that those who produce blue headed offspring are **Bb** and those who produce all red heads will be **BB** and subsequently will only produce homozygous birds for this gene.

How does this help the loft and pigeon performance?

Obviously the previous example is a very simplified version of what actually happens. For example a bird could have a dominant gene to give it a red-head and a recessive gene to give it a slightly darker shade of red head. It would be near impossible to tell the two apart. However, in the case of some genes they are not a simple one or the other affair, but are in fact additive. For example, if you mated two birds with black spots on them, their offspring may have even more spots because the two spotty genes combined. This is how performance genes work: it is obvious that birds are not either amazing racers or terrible racers, there are varying degrees. By understanding this and mating the best birds you will begin to make your birds better and better.

As far as racers are concerned, we are not interested in the colour of our birds. It is more a secondary indicator that our birds are becoming more homozygous when we have a loft full of similar looking birds. What we are interested in is the performance and health of our pigeons.

Increased inbreeding will not only increase the homozygosity of the good qualities in our birds but will also increase the prevalence of dormant health problems which will have an obviously adverse affect on their performance. This means that too much inbreeding will have the effect of an increased percentage of infertile eggs and youngsters dying young, so too much inbreeding should be avoided. Crossbreeding however will lead to an inconsistent loft of mongrels which would be impossible to breed to any system. Inbreeding is needed to get a rough idea of the genetic code of your birds so that you can better pair them together.

As a newcomer to the world of pigeon fancying you will progress much faster if your initial stock pigeons are purchased from a known homozygous family. It is crucial when breeding pigeons to have a rough idea of what the outcome will be and this is only possible with these genetically uniform of birds. You need to create a family of birds within your loft.

After keeping this family breeding regime in place for a number of years you may begin to experience inbreeding depression where, as homozygosity increases, your records show an overall decrease in performance.

At this stage it is time to introduce a carefully selected bird from an outside fancier to cross breed with one of your birds. The selected bird to cross with must be from another loft also with an obvious family with high homozygosity. The introduction of this new bird to your loft's gene pool will result in a hybrid bird that is better than both its parents. This bird will not be a consistent breeder because of its increased heterozygosity. Breeding from this bird and its offspring will, in a few generations, add new vitality to your loft and leave you with a far superior family of pigeons.

Breeding

When do I mate my pigeons?

Your fast, short distances racers should be mated at the beginning of February so that they are ready for the races in April.

Long distance, endurance racers should be mated in late February/early March to be ready for the long distance races later in the season.

Pigeons should not be raced when they have youngsters under five days old or if they've lost a good amount of their primaries. If you're planning on racing your old birds in a certain race then plan their mating around it so that they are in a fit state to perform competitively.

How do I mate my pigeons?

It's a common problem, especially among new fanciers, that two birds that you wish to mate just do not seem interested in each other or may even fight continuously.

If you force a pair to mate by confining them together in a box then they may either not mate at all, or sullenly mate against their will. Those pigeons that mated unwillingly will not rear their young as well as those happily mated pairs and will stop performing as well in races. As I have mentioned before, pigeons are a lot like us, which is perhaps why we are so fond of them. It would be unlikely that you would happily have a child with someone who you are locked in a box with, and if you did, you may well resent the child and suffer some form of psychological trauma as a result!

From experience, this is the best method of a happy mating:

DAY ONE

1. Remove all the hens and lock the cocks into the boxes you intend them to occupy. I would STRONGLY recommend, if at all possible, putting the cocks into the same nest boxes that they were in

85

the previous year. If this isn't done then sometimes they try to return to their old nest which results in territorial fighting between the cock and its new resident.

2. Put a hen into the aviary and release the cock you have chosen as a mate.

3. The pair will be very keen on each other due to being separated since December and will start their courting rituals within fifteen minutes of being put together.

4. Sometime between fifteen minutes and half an hour from the start of the courting ritual the pair can be locked back in the box from which the cock was removed. They won't fight but if the cock is acting slightly aggressively towards the hen then separate them and repeat the process the next day.

5. Repeat steps 1 - 4 for each pair of birds that you intend to mate.

6. After all required birds are paired off then release each pair one at a time, starting with the first, and give them fifteen minutes to eat and drink - do not worry if they are too preoccupied with each other to eat, just put them back into their box. I guarantee that the next day their hunger will have over-ridden their desire for each other!

DAY TWO

1. Release two pairs of birds at a time. Take care to select boxes that are not next to each other as this may result in confusion and fighting between the cocks. NOTE: Always keep a watchful eye on your mated pairs to ensure there is no fighting nor any problems with the pairings.

2. Feed the birds and allow them fifteen minutes to mate and return to their nests.

Breeding

3. Repeat steps 1 and 2 until all birds have been allowed food and water and to mate.

4. In the afternoon open up to four of the nests at once and repeat step 2.

DAY THREE

Again, allow up to four pairs of birds out at a time and feed and water them and ensure they return to their nests. They should all, by now, be doing this of their own accord. If not, don't worry, they will soon learn!

DAY FOUR

The same as DAY THREE, but this time release all the birds simultaneously. By now they should be settled with their mates and happy to return to their nests together. As always, keep an eye on them when released together until you are absolutely sure that there is no risk of fighting. It won't take more than a week until they are all happily wed!

How do I change an already mated pair?

Isolate the hen for a week, then repeat steps 3 to 6 from DAY ONE of the 'How do I mate pigeons?' section.

7

Laying

Birds that are due to start laying will soon need nesting material. Throw sawdust into the aviary about a week after first introducing the mating birds to each other, along with a bit of straw and maybe some tobacco stalks to build their nest with.

Expect your hen to lay in the afternoon ten days after mating, with a second egg being laid in the afternoon of the following day.

Give your pigeons a handful of linseed every day after the mating until the eggs are laid because the natural oils will help the egg pass through more smoothly.

Some fanciers force their hens to lay again by removing their eggs after a few days. I wouldn't recommend this because it causes a real strain on the birds; the shock of losing an egg and the stress of having to produce another can be a very large risk. It is a lot easier on the bird if you let her sit for between a week and two weeks and then remove the hen from the aviary and her nest for a week to rest. When she returns, she can mate and lay again without risk to her health.

Strengthening egg shells

Pigeons need to get the right substances from their food and drink in order to make healthy eggs with strong shells. My grandfather always insisted on making, what he termed, his 'Sturdy Egg Balls' and I don't remember him ever having an egg laid without a shell, which is why I have started using it to great success! If you do find a shell-less egg then remove it as quickly as possible from the nest and replace it with a fake egg so as to not alarm the hen. My grandfather's secret recipe:

Lime	30%
Crushed mortar	15%
Crushed house brick	15%
Crushed hemp seed or linseed	15%
Aniseed	5%
Crushed egg shell	15%
Kitchen salt	5%
Cuttlefish	One small bone, crushed

Mix it all together and add water until it becomes a thick paste. Roll the thick mixture into approximately 0.25kg balls. Dry it in the oven on a low heat for an hour (or put it in the garden if it's sunny) until dried.

Put one of the balls in with your pigeons and they will love it and lay fantastic eggs!

To be on the safe side you should always have a small amount of crushed brick, mortar and grit available to your birds so they always have enough. Not only is this used in egg laying, but it is also stored in their guts to help them to digest their food.

What if my hen is having trouble laying?

If a hen has yet to lay by the tenth day then examine her. If she is in trouble and her egg is not passing properly then soak a feather or a soft paintbrush with olive oil and lubricate her passage. Try giving the hen a laxative as this may help. If this doesn't work you have two options:

- Call a vet.

- Try to help her yourself: put a lubricated finger inside her and carefully break the egg and remove all the pieces. WARNING: if you do not remove every last piece then your hen may become seriously ill. It is always recommended that you consult with a specialist except as a very last resort.

Always replace any broken or damaged eggs with artificial eggs because the shock of losing an egg can often cause the hen to become very distressed. Sometimes it's possible to repair an egg if it's just got a minor crack.

Repairing cracked eggs

Small cracks can be repaired as follows:

1. In a darkened room, hold a torch behind the egg and locate the crack.

2. Mark the location of the crack with a soft pencil

3. Cover the crack with PVA glue and let it dry

4. Add another coat of PVA glue

If the crack is particularly bad and, from looking through the egg with a torch, you can see that the yolk and egg sack are still in perfect condition, you can place a piece of boiled eggshell over the gap and then cover with PVA glue to seal it in place.

Incubation Period

After eight days check whether or not the eggs are fertile by holding a torch behind them and checking whether the light is passing through.

Fertile eggs - appear opaque due to the growing bird within.

Unfertile eggs - will be transparent.

The cock and the hen will share the responsibility of sitting on the egg, usually with the male sitting on it for a period during the day (usually while the female goes to feed and exercise) and the female sitting on it the rest of the time.

Hatching

The young pigeon will break its own way out of the shell about seventeen days after it was laid. Sometimes it is necessary to help it by lifting off portions of shell with a fingernail, taking care to not break the interior membrane; this should enable the pigeon to free itself.

NOTE: The pigeons will not all hatch at the same time. Sometimes there will be a difference of several hours.

8

The Youngsters

The parent pigeons will start to form the nourishing milk that they feed to their young after about two weeks of sitting in the nest.

When they are seven days old you will notice that the crop (the bulge of flesh below the oesophagus, see the picture on the following page) is starting to contain a few grains. At this stage you need to put the ring onto their leg before the feet get any larger (if you find the ring hard to put on, a little olive oil used as a lubricant will help it slip on). Once the ring is on, move the new pigeons into a clean nest with clean bedding.

After this it's time for 'hands off'. Do not touch or move the pigeon in any way, not even to remove their droppings, until they are moving around on their own. They need to be left alone and kept as dry as possible. To this end I sprinkle sifted ash from my fireplace and sand around the nests to absorb excess moisture.

If your youngsters are chirping constantly, or struggling and moving around, this is a sign that something is wrong. It could be that they're ill or it could be that the older birds are not raising them properly.

The crop

If this is the case, or if they leave the nest early (before about twenty five days) and their droppings are watery, then it's not worth the time and effort to keep these birds and they should really be culled as they will never become prize winners and their parents are obviously not good breeders. I know this may sound harsh to some people but, among other things, a single weak pigeon could catch an infection making all of your birds ill.

Signs of a healthy newborn

- Releasing their droppings over the side of the nest
- A good covering of thick down
- Keeps still and does not make much noise or commotion in the nest.

Signs of a bird not developing properly

- Runny droppings found in nest
- Noisy and restless
- Abnormal growth

Schedule for weaning youngsters from the nest

20 days old

Teach the birds to feed for themselves by starting to put a small amount of food into their nests.

25 days old

Start weaning youngsters. Some people believe this is a little too young, but I have found this is the ideal age for the bird to start getting accustomed to its surroundings and it only benefits from the early start.

It is very unlikely that your bird, will know where the drinking water is, so push their head under the water so they know where to find it. If they still aren't drinking by the second day then do the same again, until they are drinking on their own.

26 days old

Put the birds onto the landing board in a cage box to get used to their surroundings. Those that spend a lot of time here will be much less likely to get lost when training begins because they will know the neighbourhood before they've even started to fly.

27 days old

They will be eating and drinking with the other birds; a distinct head start on the birds of fanciers who believe weaning should start after five or six weeks. Ensure that the older pigeons aren't preventing the young ones from getting enough food.

Check the crops of the youngsters regularly to feel that they are eating and drinking (the crop will feel swollen if they are). Again, if they are still not sure where the drinking water is then put their heads under the water every day until they learn.

9

Finding Good Stock Birds

R esearch, research, research.
There is no point buying pigeons if you don't know what kind of birds you need to race to actually stand a fighting chance in your local races. First of all, you will need to look into the race results of the races you plan to participate in. If possible get the race results for the last two or three years for the clubs and amalgamations that you will be competing in. Contacting the Royal Pigeon Association will enable you to find out who your local race secretary is; he or she will be able to help.

When you have the results, sort them by liberation point (where the birds were released from) and direction of flight for both old birds and young birds (i.e. east to west, north to south etc.) and then work out the average speed ranges of the top ten birds in the races on each course over years. If the average speed for a course is, say, over 1,600 metres per minute, then you know that for that course you will need to enter fast pigeons to win. If you see the winners have speeds of below about 1,400 metres per minute, then you know that the course is tougher and will require stronger birds to cope with the harsher weather or tougher land conditions.

Source your birds from a quality breeder

You will now have an idea of the style of racing you will need in order to win (fast and nimble pigeons or stronger and slower pigeons) and you will have to find birds to complement this.

When searching for a fancier to buy from you need to ensure you consider his:

- Reputability - Ask around with your local clubs and fanciers and find out if anyone has had experiences with buying from them. They should be willing to give you pointers regarding the racing style his birds are used to, how he trains and feeds them and how he medicates them.

- Transparency - The fancier can show you good recent race results (with proof), as well good race results as for at least the last five years. Remember to take care when a fancier only tells you how many prizes his birds have won. When presented with two fanciers, one who won ten prizes and one who won three prizes you would be forgiven for assuming the first was the best to buy from. However, if you then discover that the first fancier entered a hundred birds to win those ten prizes whilst the second entered three and won three then you should immediately see who the best to buy from is.

- Uniformity - When you look round the fancier's loft the birds should all look very similar, indicating a good breeding program producing birds with a high homozygosity. They must be willing to sell you birds from this bloodline, along with any tips on mating.

I would advise against accepting free or cheap pigeons from people without first being positive they are from a race-proven line; although these people may be trying to simply help, you will have a tougher future trying to get great offspring from bad birds.

Remember: Your MAIN goal is to get proven performance of the line you are buying from. This is the only way you will start putting money into your wallet! To this end always buy from the fancier directly rather than from someone who is buying birds and then selling them on. Individual fanciers have the reputation of themselves and their birds to consider whereas pigeon brokers are more likely to be only interested in making money out of you.

Summary

Always view the birds before you buy and look for healthy, vibrant pigeons with shiny feathers. Choose medium to small in size.

You will have to be prepared to make an investment with the initial stock. If intending to breed then you will need to buy about six birds from the same, proven bloodline and breed them as a family until you have built up your own loft of uniform pigeons, then you can start careful cross breeding to introduce new traits to your family. If you don't initially breed the birds as a family then you will undo all the work that the other fancier did and will have wasted your money. The best racers to buy will be first generation crosses between two old, long line bred or inbred families. The best breeders will be from a line bred or inbred family with a reputation for producing top racers, without suffering from inbreeding depression and needing to resort to outside crossing.

As with all aspects of pigeon racing, if you want to win, you have to do your research and never do anything unless based on information and provable results.

10

Lost and Found

R acing pigeons do sometimes get lost. They can be disorientated by weather conditions, fatigue or dehydration. All racing pigeons wear a permanent band called a 'life ring', which is put on when they are about a week old. These bands give four vital pieces of information:

- Union (organisation) abbreviation, usually indicating the country of origin
- The year of issue, either in full or abbreviated form
- A second series of letters or a single letter
- A number – up to seven digits long

These pieces of information can be used to identify the owner. Sometimes, the pigeon may have an additional plastic ring with the owner's address or phone number on it or a phone number may be ink-stamped on the underside of one of the wing feathers. The easiest way to return the pigeon is to contact the organisation nearest to you.

Once you've reported it, keep the pigeon confined until you hear from its owner. However, it isn't always necessary to trace the owner. After twenty four to fortyeight hours rest with food and water, most pigeons will find their own way home. Simply release the bird in an area free from obstacles.

Until you release it, put the bird somewhere to protect it from predators and offer it food and water. Don't risk the health of your own pigeons by putting it with them.

11

Pigeon Control

Whether you have property that you want to keep pigeon free, or you want to simply protect parts of your property from pigeons, you might be forced to consider ways of controlling pigeons' access to your property.

It's been shown again and again that killing pigeons is only a very short term solution. Culled pigeons are usually replaced within months if not weeks. The best solution is to make the surface unattractive so that pigeons don's want to roost on it. Deterrents and anti roosting products can be one hundred per cent effective if installed correctly and regularly maintained.

In the UK, the Wildlife and Countryside Act makes it an offence to place any article in such a position (deliberately or otherwise) so that it traps or injures a wild bird or animal. The Animal Welfare Act 2006 also makes it an offence to cause suffering to any animal or bird.

Most people who have a problem with pigeons call a pest control contractor. However, this might not be the best option. Many pest control contractors are ignorant of the law when it comes to wild birds. And if they get it wrong it's you – the property owner - who will be prosecuted.

Lethal control methods can only be used in the UK if all non lethal methods have been tried and have failed and where a demonstrable health and safety risk exists. It's illegal to kill birds because they are soiling or causing damage to a building. Lethal control has been proven to increase pigeon flock size. Pigeon numbers increase back to the pre-cull figure within a few weeks of a cull. Most anti-roosting products can be 100% effective so long as they're used correctly.

Pigeon pill

The contraceptive pill has been trialled on wild pigeon flocks in Europe, but it isn't licensed for use in the UK. A pharmaceutical company has introduced a contraceptive called OvoControl P for pigeons in the USA. This use of this drug has caused a great deal of controversy, mainly because its active ingredient (nicarbazin) is commonly used to control coccidiosis. The high levels of nicarbazin that pigeons intake if given OvoControl P causes them to lose any natural resistance to coccidiosis.

Lethal methods

Poison

Although poison was extensively used in the UK to control pigeons in the 1960s and 70s, the use of poison bait is now strictly controlled. A licence is needed and these are now rarely granted.

Shooting

Shooting is widely used by pest control contractors. Birds are usually shot at night in their roosting places. Many of the birds shot are simply wounded and left to die.

Pigeon Control

Cage trapping

This involves encouraging pigeons into a baited trap placed in their roosting or feeding area. The birds are then removed and killed. Although some pest control contractors may state that they'll release the birds, it's much more likely that they will be killed as otherwise they would fly straight back to the site. It's also an inefficient method as once traps have been in place for two weeks, the pigeons will avoid them.

Bird of prey

Using a bird of prey to scare pigeons is becoming more common. Many see it as a more natural method of control. However, using one species of bird to kill another when the bird isn't the pigeon's natural predator isn't natural. Also, other species of birds might be killed, including protected species. This is an expensive and controversial method of control, although a number of companies offer this service.

Non lethal methods

Sonic bird scarers

Some of these devices also include visual deterrents such as flashing lights. There are three main types:

- A sonic system produces a noise such as a siren or a loud bang
- An ultrasonic system produces sounds outside the range of the human ear
- A bio-acoustic system uses up a mixture of methods:
    - Predator calls that mimic the sound of a predatory bird

- o Alarm calls that mimic the sound of the target species makes when being attacked or in danger
- o Distress calls made by the target species

Noise based systems tend to be ineffective. The birds quickly get used to sonic noise, ultrasound and bio-acoustic techniques. They have no effect on breeding birds that have eggs or young in a nest. Despite being expensive, these devices are particularly ineffective with pigeons as pigeons are one of the few birds that have no natural distress call.

Repellent gels

These products have a short lifespan and can cause a lot of damage to the surface they placed on. Often, they only remain effective for a matter of hours. As many pest control contractors fail to apply sealant, birds can become glued to perching areas. If the gel gets on their feathers, when the bird attempts to fly it plunges to its death.

Electric shock systems

Although common in the USA and also used in some European countries, electric shock systems are not legal for use in the UK. The device consists of thin steel wires that are enclosed in a track attached to the surface to be protected. A low current of electricity gives any bird landing or walking on the surface an electric shock.

Anti roosting spikes

These are designed to physically prevent a bird from landing on ledges. Simple to install, they are very effective and durable. Anti roosting spikes are usually installed by using silicone gel, which doesn't mark or damage the surface. The spikes can be easily removed if necessary.

Pigeon Control

Nylon bird netting

This is usually installed to prevent a bird from gaining access to an area rather than a specific perch. It tends to be expensive and performs poorly. It's essential to regularly maintain and re-tension the netting as it expands and contracts with changes of temperature. Birds frequently become trapped behind the netting. It can be effective on small areas if it's well looked after.

Post and wire systems

A series of vertical steel posts are installed on a flat surface and spanned by thin steel wires a few centimetres above the surface to be protected. The steel wire is joined to the posts by small steel springs. When a pigeon tries to land on the protected surface, its feet first touch the wires, which move and make the bird feel that its landing is unsafe. The system is extremely expensive and the wires often snap or become disconnected. Pigeons often nest behind the spring wires because they stop nests and their contents from falling to the ground below.

12

The Homing Instinct

A lthough in ancient times the flight of ravens, crows, cranes and owls was observed, only the pigeon had a strong enough homing instinct to be relied on to return without fail.

To pigeon keepers, it seems that their pigeons have something like an invisible elastic band connecting them to their home and drawing them back toward it. When they are taken elsewhere, this band is stretched. If when returning they overshoot their home, this connection pulls them back again.

Numerous experiments have been carried out to ascertain how pigeons home. So far, research has not uncovered exactly how their navigational ability works. In general terms, pigeons can be relied on to find their way back, although from time to time they do get lost when disorientated by weather conditions, fatigue or dehydration.

The inertial navigation hypothesis (that pigeons register the movements of the outward journey) is no longer taken seriously by researchers.

It's clear that visual cues aren't essential. The pigeon's navigational system is largely non visual. On sunny days, the pigeon's sun compass can play a part in their general sense of direction. But that isn't enough to explain their ability to find their way home.

One ability suggested is the pigeon's reaction to infrasound. Laboratory experiments show that pigeons are unusually sensitive to low-frequency sounds. However, this does not prove that they can hear their homes from hundreds of miles away and there is no evidence to suggest that infrasound plays a role in pigeons' homing ability.

Research has shown that smell can sometimes be important. Pigeons create odour maps of their home location and use these to orient themselves. In some circumstances, the sense of smell plays a part in the orientation of pigeons, but it can't by itself explain how pigeons find their homes.

The hypothesis that pigeons react to magnetic fields in order to find their way is popular, but not backed up by science. The magnetic sensitivity of pigeons has been tested in numerous experiments and the majority of the published results have failed to show any significant effects of magnetic fields.

The current view is that pigeon homing depends on a complex series of systems that involve subtle combinations of mechanisms, such as a sun compass, smell, and magnetism, or that pigeons use a single type of information that is, as yet, undefined, but scan it with several sensory systems.

Scientists are still fascinated by the pigeon's homing ability and continue to search for its secret.

13

Racing

Simply put, racing pigeons are taken from their lofts and must race home. The time taken and distance are recorded and the fastest bird is the winner.

The pigeon racing season in the UK begins at the end of April, and continues until mid September. Each pigeon organisation arranges about twelve races for old birds and about eight for young birds. The races commence with one from a distance of about sixty miles, which is gradually increased as the season progresses. All the race points are in the same direction from the organising club's headquarters, so that the birds fly in the same direction each week, travelling over the same corridor of land.

To begin racing, you will need to join your local pigeon flying club. The club will make you a member of the appropriate homing union. The unions issue identification leg rings to clubs so that every stray bird can be traced to its owner. No pigeon can be entered in a race without an official registered ring.

Each club usually belongs to a federation of twenty or thirty clubs. The federation arranges the transportation of the birds to the race points. A convoyer is responsible for feeding and watering the birds en route and liberating them at the race point.

For you to be able to compete, the longitude and latitude of your loft has to be precisely determined. Once the locations of the loft and the liberation point are known, the distance between them can be measured. You will also need to have a timing clock of an approved make. Although most fanciers have their own, clocks can also be hired. The clocks need to be checked and test run before the start of each racing season. The evening before a race all members' clocks are set and checked and then sealed.

Competing pigeons are entered into the race through the hosting organisation and released at a predetermined time and place. Sometimes there are two divisions of races - one for young birds and another for older racing pigeons.

With the traditional method of timing, the ring's serial number is recorded and the clock is set. When your pigeon returns, its ring is placed in a slot in the clock and the time recorded as the official time that it arrived home. This gives an average speed so the winner can be found.

Most pigeon clocks can accommodate ten or twelve rings in numbered receptacles so that it's possible to work out in which order the birds have been clocked. The clock is taken to the club's headquarters and the rings removed.

However, as the official time isn't the actual time that the pigeon arrives, vitals seconds can be lost when placing the ring in the clock. Also, some racing pigeons may be reluctant to have the ring removed and so reluctant to enter the loft.

The electronic timing system records automatically. The bird is fitted with a band that contains a tiny chip that can be read by a scanner. When the bird comes home, the pad or antenna at the loft entrance reads the chip and records the time of arrival. With this system, it isn't necessary for the trainer to be present.

Money prizes are usually awarded to the first three pigeons on the race sheet. Most fanciers want to win a national race. Thousands of birds compete in these races for large sums of money.

Scheduling and managing your training programme will give your birds the best chance of success. They need to have the confidence to make their own decisions and not be tempted to follow a potentially inexperienced flock of under trained birds. There is no point in entering a pigeon in a race unless it is physically fit and if a bird spends a night or two away, it needs to be given plenty of time to recover before being raced again.

It's best to organise your race entries so that each young bird gets three or four races. If you send them to every race, it can take too much out of them. These birds are your investment for future races.

You can focus on training the birds you plan to enter for a race during the week before. Leave at home the pigeons that won't be going to the race. Ideally, your birds should see the last twenty miles once or twice, a day or two before they are due to fly. They will be toned up and ready while those that remain at home won't become stale from over-training and be risked unnecessarily.

Don't place too much emphasis on young birds' performances. Just because they fly well as a youngster, they won't necessarily win prizes as a yearling or old bird. Even a young pigeon that offers only a moderate performance when young can mature into a champion. Even if they fail to return home the day of the race, they can make a good racing pigeon as it has learned to work on its own and gained valuable experience. Young birds have less incentive to return than old birds as they may have eggs to incubate or youngsters to feed.

Youngsters from a third round off a good pair will usually be hatched in June and consequently will he too young to be trained with the first and second round, but in my experience, they are definitely worth all the extra effort. Supposing the fancier has six third round youngsters off his three best pairs, he should begin training these as soon as they are ten weeks old, which will be about the second week in August. They should be trained by themselves, just as their older brothers and sisters were trained a month or so before.

Between races train only those that are being entered for the next race. One toss of thirty miles for those that raced the previous week. Two tosses of twenty and thirty miles for those that have not raced for two or three weeks.

Racing around the world

There are racing pigeon clubs all over the world. Pigeon racing is the national sport of Belgium and the Queen of England has a loft of pigeons. Racing pigeons in America is also becoming a big sport. National organisations, as well as local clubs, exist in most countries.

Most breeds of pigeon have speciality clubs. These include groups for breeds such as the Tippler (known for endurance flying), the Birmingham Roller (an acrobatic flyer which performs rapid backward somersaults), and the Racing Homer (a bird which can return rapidly from distances in excess of 700 kilometres).

Asia

Pigeon racing is popular in parts of China, Pakistan, the Philippines, Japan and Taiwan, where millions of dollars are bet on the races. It's becoming increasingly popular in India - especially in Chennai, Bangalore, Hyderabad, Kolkatta and Tuticorin – and Taiwan has more racing pigeon events than any other country in the world. Half a million people race pigeons on the island.

Australia

The largest racing organisation in Australia is the Central Cumberland Federation in Brisbane. The state of Queensland also has a number of clubs and organisations. The Queensland Racing Pigeon Federation organises annual pigeon races starting at about 145 km up to over 1,000 km. Australia's Premier One Loft Event is the Mallee Classic held in Ballarat Victoria.
Central Cumberland Racing Pigeon Federation
http://www.pigeonrace.net/?ccf

Queensland Racing Pigeon Federation Inc
(QRPF)http://www.qrpf.com/

Belgium

The Janssen Brothers (Louis, Charel, Arjaan and Sjef) are
a famous and very successful pigeon racing family from
Arendonk, Belgium and descendants of their pigeons can
be found racing all around the world.
Koninklijke Belgische Duivenliefhebbersbond
http://www.kbdb.be/
Royale Federation Colombophile Belge, 39, Rue de
Livourne, Bruxelles, Belgium. Email: national@rfcb.be

Canada

CRPU - The Canadian Racing Pigeon Union
http://www.crpu.ca/

New Zealand

Pigeon Racing New Zealand
http://www.prnz.org.nz/

France

Federation Nationale Des Societies Colombophile de
France
http://www.colombophiliefr.com/Egares/Egares-e.htm
54 Boulevard Carnot, 59042 Lille, Cedex, France.

Germany

Nationale Sporttauvenzuchter Vereingung in der
Deutschen,
Demokratischen Rebublik Federation Colombophile
Nationale,
1071 Berlin, Wicherstrasse 10, Germany.
Verband Deutscher Brieftaubenliebhaber E.V. Essen
(Germany), Schonleinstrsse, 43, PO BOX 1792, Germany.

Malta

Federation of pigeon Racing Clubs Malta, Railway Avenue, Hamrum, Malta

Netherlands

Nederlands PostDuivenhouders Organisatie
http://www.npo.nl/site

Portugal
Federacao Portuguesa de Colombophile Rua Padre, Estevao Cabral No. 79, Sala 205, 3000 Combra, Portugal.

Romania

Romania is one of Europe's hot spots for pigeon racing. Many pigeon breeders join the National Association every year, triggering more and more competitive challenges.
Romanian Racing Pigeons
http://www.rrp.ro/

South Africa

The richest One-Loft Race existing is found in South Africa. The Sun City Million Dollar Pigeon Race involves 4,300 racing pigeons from 25 countries that compete for their share of the $1.3m prize.
Western Cape Pigeon Racing
http://www.westerncapepigeonracing.co.za/

Spain

Royale Federation Colombophile, Eloy Gonzalo n° 24 Madrid, Spain.

Racing

United Kingdom

The first regular pigeon races in the UK started in 1881. The royal family first became involved with pigeon racing in 1886 when King Leopold II of Belgium gifted them breeding stock. The National Flying Club is a British pigeon racing club, and open to anyone in England and Wales.

National Pigeon Association
www.nationalpigeonassociation.co.uk/
The Association can supply a list of affiliated clubs and societies. It is the governing body of Fancy Pigeons and controls the issue of rings, championship Shows and pigeon exhibiting. It caters for over 200 varieties of pigeons.

Pigeon racing in the UK is regulated by six independent organisations:

Irish Homing Union
38 Ballynhatty Road, Belfast. BT8 8LE

North of England Homing Union
58 Ennerdale Road, Walker Dene, Newcastle-upon-Tyne. NE6 4DG

North West Homing Union
279 Mossy Lea Rd, Wrightington, Nr. Wigan. WN6 9RN
nwhu@blueyonder.co.uk

Royal Pigeon Racing Association
http://www.rpra.org/

Scottish Homing Union
www.shuonline.co.uk

Welsh Homing Pigeon Union
Old Timothy's Yard, Llanfoist Street, Ton Pentre,

Rhondda. CF41 7EE
Email: gail@whpu.org.uk

USA

The sport was introduced into the United States about 1875 and pigeon racing began in 1878. There are 15,000 registered lofts in the US. Pigeon racing is banned in Chicago.

American Racing Pigeon Union
http://www.pigeon.org/info.htm

Links to clubs:

International Federation-USA http://www.ifpigeon.com/
American Homing Pigeon Fanciers,
http://www.ifpigeon.com/pigeons.html

14

Training Young Birds

The most important thing is to ensure that your birds are tame and eating out of the palm of your hand. If your birds are still reluctant to be near you then you will lose valuable seconds trying to trap them when they arrive back to your loft after a flight or race.

Ensure you spend a lot of time in view of your pigeons and get them used to a regular routine that revolves around you. This means:

- A fixed training and feeding routine.

- Always entering and leaving the aviary at the same time and via the same entrance.

- Speaking to your pigeons whenever they are within earshot.

After they've been weaned

Training begins from the day the birds are weaned from their parents, at about twenty seven days old.

Put your birds into a cage on the landing board of the loft for a few hours a day before they can fly. Make sure the cage is built so that they can see all around them. They will start to get to know their surroundings and you will avoid fly aways when they start their flight training.

It is best for young birds to have constant access to food and water until they are feeding themselves with confidence. At this point the young birds move on to being fed twice daily, just like the old birds.

When they begin flying

After about a week after leaving their parents, at about 5 weeks old, you will notice the youngsters starting to fly up onto their perches to roost for the night.

At this stage you can either place them in an open cage on the landing board as before or leave the trap door open for an hour a day and let them make their own way out when curiosity gets the better of them. It is important that you never let these young birds out when older birds are flying outside; these old birds may flock and fly further than the youngsters can manage.

It is now that they should also be trained to enter through the trap. Let them out as normal and then close the trap behind them, making this the only way for them to return inside. Young birds can be stubborn at first but eventually hunger will get the better of them and they will use the trap. Just be patient and don't force them because this will make them even less keen.

Do not feed them before letting them out. If they do go out you want them to return for food when they're hungry. Make your feeding call every time you are about to put food down for them, then when they start feeding, watch till the first bird has stopped feeding and remove all the food. The birds that didn't return quickly when you called will not have time to eat properly.

They will have to wait until the next feeding time to get more. All your pigeons will quickly learn to come as soon as you call.

When the birds start flying and flocking together it is time to start letting them out for exercise before every meal and reinforcing the return for food mentality in them.

Tip: if you find that your pigeons are too excitable when you let them out, increase the amount of legumes (peas, beans etc.) in their diet. This will calm them considerably.

Basket training

Basketing is the name given to putting your birds into a specialist transportation basket. It is a crucial part of racing so it is important to get your birds comfortable with it as soon as possible. This can be done in a number of ways:

- Store an open basket in the young bird loft so that they get used to seeing it and can explore it.

- Keep young birds in a basket overnight in the loft. They will learn to feed and drink from the containers attached and not worry when they are basketed for training or racing.

- Put the pigeons into a basket when leaving your birds out on the landing loft before they can fly.

Don't handle your birds roughly as they will become afraid of you and will fly wildly to escape.

Youngsters are ready for training when they are exercising happily and easily in the mornings and evenings and have grown accustomed to the being locked in the baskets.

Training flight

If any of your birds are still unhappy with the baskets or seem unhealthy then do not risk flying them. It is better to be sure that your birds are healthy enough to fly home than to risk flying them because you were impatient.

Before training begins the birds need to be wing stamped (Association rules) and vaccinated against paramyxovirus (legal requirement).

Pick a clear day with few clouds. Don't train your pigeons in an east wind, when there are heavy clouds about or when thunder is likely. If you can't see the sun, do not take your birds training. If you should have taken them to the release point, and on arrival the sky has clouded over, either wait until the clouds disperse or take your birds home.

Young birds do better being trained in the morning. Give the birds a practice toss from the other end of your garden, in sight of their loft, just to get them used to being basketed and released. From here they can start to be trained further afield.

Their first proper training release should be done at around five kilometres from the loft and should go as follows:

1. Start at mid morning to give the pigeons plenty of time to return home before nightfall.

2. Feed them half of their normal rations and don't give then any morning exercise. By limiting the amount of food they eat you will increase their desire to get home.

3. The weather should be clear with a light wind and minimal cloud cover.

4. At the liberation point remove the basket from your car and put it on the floor. Leave it for around fifteen minutes to let your birds calm down and get their bearings. This is a very important step, as birds released as soon as the basket is put on the floor will be scared and excited and will fly off without a clue where they are.

5. Ensure that back at the loft the birds must enter through the trap.

As your pigeons get more comfortable with being outside, they will venture further away. When they start circling in the air and flying away for short periods, you can begin to let them loose at different distances to learn to find their way home. Be aware that not all birds return when let out to fly.

Once your pigeons know where their loft is, and how to enter, you need to teach them the other ground rules. They should only sit on your property, preferably only on your loft. If they start to perch on your neighbour's house or telephone wires, throw a tennis ball or similar at them so that they learn that that place is off limits.

On their first few training flights you will notice that it will take your birds longer than expected to return home. If you have followed my instructions then you need not worry, they will return faster each time and the amount of birds going missing will be minimal (you will have to accept, however, that sometimes birds can get lost and not return, unfortunately).

Keep tossing your birds from this same point until they start arriving back at the loft in good time; at this point increase the liberation distance by ten kilometres. When they return in good time from this new distance then add another ten kilometres and so on and so forth.

When the birds are being trained from twenty kilometres, I start to introduce them to bad weather by taking them out regardless of the elements (aside from hurricane force winds!) because I think that it is a mistake to only train birds to fly in the sun. If they're going to meet all sorts of weather in a race situation then they should have at least seen something similar to it when training!

If your birds get confused and lost, and take several days to return, give them a few days to rest and then try again by taking them back to the last place from which they returned home in good time. Remember that although pigeons are born with their homing ability, it's an ability to find where they want to be, which may not be where you want them to be. So it's important that your place is warm and welcoming.

It can be useful to have someone at home watching your pigeons return to see if they come back in the order that you release them. This way, you can see which of your birds are training the best.

As soon as your birds return home, whistle them inside the loft. They need to be trained to enter the loft as soon as possible on their return as only seconds can separate the winner from the next pigeon.

Most shorter races are won in the last few miles from home, so it's essential for your birds to learn the most direct and shortest route home. The more often your birds fly the few miles of the route to your loft, the better they will get to know it.

During training you will lose pigeons that aren't smart enough to find their way home as well as those that are not strong enough or don't have the courage or will to come home. Young racing pigeons take about six months to train before you can use them to compete in a race. A pigeon that survives the hazards of racing could compete from about six months of age and be in competition for ten years. However, the average career for a racing pigeon is usually three years.

Although training methods can vary, the most important thing is to be consistent in whatever method you use. Always try to feed and exercise your pigeons at the same time of day and follow a routine.

15

Training Old Birds

Y ou can't let pigeons out to fly at your home that were not hatched there. If you buy a pair of adult pigeons, you will not be able to let these birds out to fly as they will simply return to the loft they were trained to home to (if they were trained). Therefore, you need to breed a pair of pigeons and train their babies.

If this is your first year having pigeons you will have young birds that have just completed their first race season and yearlings that you bought for breeding.

These stock birds would not have had any training even though they will be exercising outside around the loft. It is not a good idea to race all these birds in your racing team because, as potentially valuable breeders, you can't risk losing a high percentage through poor training.

If you chose to race them, then experience for them is key: race them in the shorter races but don't participate in the longer distances at this stage because you can't be sure how they will fare.

It is very important to keep good, accurate records of all training and race results for your older birds. It may not be until a pigeon is two years old that it really starts to shine as a racer or you find the perfect pairing for it to be a great breeder. However, if results are consistently bad for a bird, i.e. frequently placing in the bottom half of every race and training toss then it should be removed from the loft.

Race training

Before training starts, the birds will be sitting on their second batch of eggs. Replace these with artificial eggs after ten days to stop your racers from wasting energy on rearing more youngsters and to delay the moult (which starts after the second batch of eggs hatch).

Two weeks before the start of training your birds should already be outside exercising happily.

If your birds are sitting on eggs then they may be reluctant to go and exercise at first. Aim to have them flying for a minimum of thirty minutes for five days then increase their flying time until they're flying for about an hour. It is at this point that allowing cocks and hens to fly together is an advantage because they will happily stay out longer if around their mate.

If your birds are exercising well from the loft then you don't really need to train more than twice a week. When I say 'exercising well' I mean that they are eager to leave the loft when you go to let them out and, once released, they fly hard and fast away from the loft and come back looking tired. If they are just flying in circles around the loft then this is not really good enough and the amount of training tosses will have to be increased.

The training tosses for Old Birds should start at around ten kilometres and work up to about sixty kilometres but, unlike the youngsters, they do not need so many multiple releases from the same point as they will already know exactly where their home loft is.

Best times to race cocks

- When he's been sitting for five days
- When feeding a youngster just before his second round of mating
- When driving his hen

Best times to race hens

- Any time when they're sitting on eggs. The closer her eggs are to hatching, the more she'll want to get home.

If the hen is sat on artificial eggs, then remove an egg that's cracking, and soon to hatch, from another nest and place it under her before a race. This will trick her into thinking that she has to get back urgently in order to tend to her young. Return the egg to its real mother as soon as the racing bird has been basketed to ensure that the egg doesn't get damaged.

16

Record Keeping

This is the main difference between a great pigeon fancier and an underachieving pigeon fancier.

If you want to breed the best birds, know which birds are worth keeping or find your winning system then you will have to set up some sort of record keeping structure. There is no right or wrong way to this but it is best to be organised and keep as much information as you can. If you use the record templates supplied with this book then you will have a whole range of very useful data to make conclusions from.

How to best utilise your records for SUCCESS

Every pigeon fancier can understand the following simple study: some fanciers will claim that the only way to win races is to train their flocks over fifty kilometres twice a day, while some very successful fliers say they train their birds as little as six times a season. Instead of testing which of their two systems works best, they stop there and remain convinced that their own is the best method. The only evidence they each have for their individual methods may only be that one man says he trained less one season and his results weren't as good, whereas one says he tried training more and his results weren't as good.

This leads us to the next point that any real scientist must consider when performing their tests: every detail of comparison should be identical except the one being tested.

Let's take the example of a driver who wants to discover which of his two cars was the fastest. Would he race one during gale force winds and heavy rain and the other on a dry, sunny day then conclude that the one raced on the dry sunny day was the fastest? No, because they were both raced under different conditions and it was obviously not a fair test.

This example should apply particularly to pigeon racing. If a fancier compares results from one year with a previous year then he can't possibly make accurate conclusions as too many variables will have changed. One year the pigeons may be badly infected with parasites (perhaps intestinal worms or pigeon fleas) and then not the next, or the feed or the grit being given to the birds may have changed.

No test should be made except when ALL the conditions but ONE are the same.

Has any fancier scientifically attempted to test his idea about the need for frequent training? Has he ever trained half his flock severely and the other half moderately?

Split testing

Split testing is an amazingly powerful process that, if used properly, will allow you to make conclusions that will dramatically increase the overall quality of your birds. In order for any testing you do to be statistically relevant (i.e. that the result wasn't just down to chance) you need around sixteen pigeons per test group. This means that the best situation would be where you had a total of thirty two pigeons or more, to test. Any testing is better than no testing and by testing over a longer period you will still get useful results.

How to accurately split test your loft

Split your loft equally and RANDOMLY into two halves.

It is important to not split them yourself because we, as humans, are completely unable to ever split anything truly randomly. Even by trying to do something at random, we are affecting our results by TRYING.

To do this:

- Assign each of your pigeons a number starting at 1 and going up to however many pigeons you have. If you've got 30 pigeons then give each pigeon a number from 1 to 30.

- Use the free random number generator at www.random.org/sequences

- Enter '1' as your smallest value and the number of pigeons you have as the largest value. Enter '2' for the number of columns.

Total number of pigeons taking part in the test

RANDOM.ORG

Random Sequence Generator

This form allows you to generate randomized sequences of integers. The randomness comes from atmospheric noise, which for many purposes is better than the pseudo-random number algorithms typically used in computer programs.

Part 1: Sequence Boundaries

Smallest value 1   (limit -1,000,000,000)

Largest value 30   (limit +1,000,000,000)

Format in 2   column(s)

Total number of pigeons taking part in the test

The length of the sequence (the largest minus the smallest value plus 1) can be no greater than 10,000.

Part 2: Go!

Be patient! It may take a little while to generate your sequence...

Get Sequence   Reset Form   Switch to Advanced Mode

How to fill out the Random.org/sequences form

- Press 'Get Sequence'

RANDOM.ORG

Random Sequence Generator

Here is your sequence:

18	7
25	3
28	26
21	29
4	23
30	2
5	9
27	12
1	11
20	17
24	16
10	19
13	6
14	8
22	15

Timestamp: 2010-05-10 11:33:33 UTC

Example results from the random split produced by Random.org/sequences

- Split your birds into two groups based on the random numbers produced: all birds with a number in the left column go into one group; all birds with a number appearing in the right column go into the other.

- Now that you have two completely random groups you can begin setting up the testing.

- Pick ONE part of your routine that you wish you test. Let's say you decide to test your feed mixes.

- One of your two pigeon groups will be your control group, the one that you will feed as normal, and one will be the test group that you will feed differently.

- Keep them on this routine for a year and at the end, analyse their race results (using the forms contained with this book). You will then know that the best thing to feed your birds is what you fed your best performing half of pigeons.

Obviously, the longer you run your tests for, the more conclusive your results will be.

17

My Best Systems

The darkness system

This system works very well when applied only to your young hens, and not to the cocks. Leaving the cocks to mature normally and progress through a normal moult ensures they will be ready to race in the following season while your hens are resting and breeding.

This is a method for quickly raising young birds and gets amazing results when used properly. It revolves around tricking your pigeons biological clock into thinking that winter is coming when it's the racing season. They moult their body feathers very quickly but keep their flight and wing feathers; preparing for winter and conserving energy. They quickly become adults in all respects except the adult flight feathers.

A lot of fanciers have a number of misconceptions about this system: they will tell you that these pigeons are terrible racers as old birds. I believe this is down to people over-training their youngsters and have not seen any evidence of any fall in performance as the years progress.

Advantages

- Young pigeons quickly get adult immunity to a large amount of illnesses and diseases that they would normally fall prey to.

- With no wing moulting, the pigeons can fly without the pain and worry of losing feathers.

- The lack of moult means they can race the entire season if their health allows it.

- They become sexually mature quickly meaning that this system can be used in conjunction with the *widowhood system.*

Disadvantages

- Potential losses from birds if not raised as laid out in this book; your pigeons will be ready to fly earlier than normal and, without proper training from a young age, will not have adequate knowledge of the area surrounding their loft.

- You must install louvre panels over ventilation spaces to ensure adequate ventilation whilst simultaneously keeping the light out.

When do I implement this?

As soon as your youngsters are no longer reliant on their parents for food which will be some time around May.

How do I implement this?

- Keep your loft light and allow your birds to see natural sunrise in the morning. After nine hours of light cover the windows and darken the loft to a level slightly darker than dusk. It is important to keep to a rigid routine for the birds to adapt.

- Increase the amount of protein your birds are receiving by increasing the amount of wheat in their diet by 50%. They will need this for their rapid growth.

- Basket your birds and put them outside on the landing board of the loft for two hours in the morning and two hours in the afternoon each day.

- When you notice that the birds have started flying up to their perches in the loft, it is time to begin letting them onto the landing board without the cage.

- For an hour a day put the youngsters onto the outside board and they will gradually start taking to the sky and perching on nearby trees and roofs. Do not let them out unattended because at this stage they are very vulnerable because they are not yet good enough to out-fly predators.

- As soon as the young birds start to fly together as a flock you should begin training.

- The training should last for three weeks and be almost continual, weather permitting. Start with training tosses from the end of your garden, slowly working up in increments to 5km and then over the weeks to around 60km. If you know the direction your birds will be travelling when returning from a race then my advice is to make all these tosses in a line "as the crow flies" towards the rough start point of the race. This is more important for the young birds you've singled out for participation in one or two long distance races.

- After the three weeks training let your birds rest and relax. It should be at this point that you are starting to make your selections as to which birds you will enter in the long distance races and which birds you will enter in the short distance.

- Put the youngsters back onto normal daylight hours four weeks before the first race and separate the cocks from the hens. As I said previously, it will be the hens that we'll race whilst the cocks will be saved until next season. Give them their daily flights around the loft but do not give them any strenuous training.

- Start training the hens again after two weeks, again slowly increasing the distances they're tossed from.

- Feed the birds sparingly at the beginning of the week before a race and gradually increase the amount of food they receive each day. Their food rations should increase so that they are fed the full amount on the day before the race and a slightly lowered amount on race day. This may seem counter intuitive but pigeons can go for days without food and restricting their diet for a week seems to really make a difference.

- Make a record of exactly how much the birds are fed on the run up to race day. This will enable you to be certain this method works.

- After three weeks of being back on normal daylight hours the birds will be taking part in their first race. It is now that you start the birds on the *widowhood system*. This system will be applied to both the young birds raised with the *darkness system* and the older birds.

Widowhood system

This is basically just a complicated name for something which is, fundamentally, very simple. It can be used as part of the darkness system or on its own.

It revolves around separating the sexes from each other and only letting the mates be together for between fifteen and thirty minutes before a race and one or two hours after a race. By racing the hen she will be eager to get back as soon as possible to see her mate and to sit on her eggs, making her race faster.

How to implement this

- Do not let the cocks and hens have any interaction whilst on this system. That means letting them out for their exercise at different times and training them separately.

- Let the pairs be together for around half an hour before the bird you have selected to race is basketed, ready for travel.

- After the race let them stay together in the nest for between an hour and two hours.

- Two weeks into the racing season start using your UV lamps in the loft to extend their days to between fifteen and sixteen hours of daylight. This is the amount of daylight they would receive in the middle of summer when the moult is very slow; they will lose one or two of their primary feathers but will still have the vast majority at the end of the season.

- After the race season stop the darkness system and allow them to enjoy natural sunlight. They will start to moult heavily on their wings and their bodies. These birds will finish the moult the year after, which is why it is advisable not to use this system for the cocks.

- The hens can spend the next year raising young and finishing their moults whilst the old cocks will have a normal timed moult and be placed onto the *widowhood system.*

Jealousy system

If a cock returns from a race and his partner has started the courtship ritual with another cock then the two male birds will fight. Remove the hen's new partner and basket

him where he can see the old cock and the hen he had been courting. Allow the two cocks to be in the same loft and to fight for a couple of minutes before the new cock is sent off to compete. When he is let out of his basket at the liberation point he will fly as fast as possible to get back and fight for his new partner.

Glossary of Pigeon Terms

Air sacs - Nine hollow areas extend throughout the pigeon's body. Air flows through this system of interconnected sacs almost like blood in the circulatory system.

Arm - The humerus is the bone in the wing which projects directly from the body.

Ash - A racing homer of a light tannish colour over its entire body and having no bars across its wings.

Bars - Colour bands across the back part of the top surface of the wing.

Bib - A colour pattern of the front part of the neck.

Billing - Pigeon kissing. When the female sticks her bill down the male's throat and takes an offering of regurgitated food. Known as a prelude to mating.

Bloom - A white, powder-like dust found in the feathers of pigeons.

Blue Bar - A racing homer of a light bluish grey colour with two black bars across the back part of the top surface of its wing.

Blue Check A Racing Homer of a light bluish grey colour with black checked patterns on the top surface of its wing.

Bob - The part of the trap that the pigeon pushes against to get into the loft.

Bowing - Courtship behaviour of a male pigeon.

Breaking point - A theoretical point where your bird

must break from the flock of racing birds in order for it to win a race.

Breast - Crop region of the body; also includes the pectoral muscles.

Bull eye - A very dark coloured eye. A thick iris with "breeding grooves" which are black or dark in colour

Cere
Beak cere The fleshy part above the nares on the upper bill.
Eye cere Bare skin around the eyelid.

Checker - The triangular blotches of colour across the lighter wing shield.

Circling - The act of a Racing Homer flying in a circular pattern around its loft

Clutch - A name given to a set of eggs laid.

Cock - A male pigeon.

Cover feathers - The cover feathers make up most of the wing and are attached to the upper part of the wing.

Coverts - The small feathers of the wings and tails.

Crest - Reversed feathers on the back of the head.

Crop - The first stomach of the bird for storing feed located in a fleshy neck pocket. Food is stored here for about twelve hours before being passed into the stomach and intestines.

Cross breeding - The mating of birds with no relationship within the previous five generations or unrelated birds.

Cull - A pigeon with undesirable traits.

Dam - Mother homing pigeon.

Driving - The behaviour of the cock bird for the few days before his hen lays her first egg of the nesting cycle. He forces her away from the proximity of other cocks and in a loft situation often forces her back to the nesting site.

Drop - During a pigeon race, birds will usually not arrive at the loft at the same time. Instead, they will come in groups. The first "drop" is the group of pigeons that come home to the loft in the first grouping.

Dummy eggs - Wooden or plastic eggs designed to use as a substitute for the real thing. Used to slow a pair's production.

Egg bound - A hen unable to lay a completely formed egg.

Feeder pair - A pair of pigeons used as foster parents to another pair's eggs.

Feather merchant - An individual that puts the love of money before the best interest's of the hobby or those in the hobby.

Flight or fly pen - An extension of a loft or enclosure used for exercise, normally made of wire to allow sun and restricted flight for the lofts inmates. An aviary.

Fret marks - Any horizontal mark on deformity on the feathers.

Frill - Line of reversed feathers on the neck or crop.

Genetics pair - A pair of pigeons that have been paired to propagate a particular genetic trait not normally found in that particular breed.

Girth - Circumference of the pigeon's body.

Grizzle - The colour pattern of individual feathers of a pigeon that have two colours.

Hen - A female pigeon.

Keel - The breastbone.

Kit - In rollers, it is a group of pigeons numbering 15 - 25 birds.

Kit box - A rabbit hutch like cage that is used to house a kit or team of rollers.

Landing board - A large flat surface on which the birds land before entering the trap and the loft.

Mandible - The beak or bill.

Mealy - A racing homer of a light tannish colour with red-brown bars across the back part of the top surface of its wings.

Milk - The cottage-cheese looking crop substance produced by both cock and hen to feed the young from hatch till about ten days.

Moult - The yearly process of losing all feathers and growing new ones in a systematic manner.

Muffs - Large feathers completely covering the legs (tarsi) and the toes.

Natural system - A racing system in which the birds are kept paired, sitting on eggs or rearing youngsters and used for motivation during a race.

Nest box - A box that the birds nest and raise their young.

Old bird - Any racing homer more than one year old.

One loft race - A specific loft where breeders send their birds and all the birds are raced from that one loft.

Pectorals - The large muscles lying on both sides of the keel.

Pied - A racing homer with white feathers on its neck or head.

Pin feather - A growing feather of the young pigeons that has not yet broken through the shaft.

Pipping - The process of the young bird chipping out of the egg shell during hatching.

Plumage - General feathering

Primaries - Last 10 large flight feathers in the racing homer's wing.

Pumpers - Pigeons used as foster parents.

Rolling - A continuous downward performance of backward somersaulting in flight.

Secondaries - The inner flight feathers of the wing which provide lift.

Show bird - One that is as close an example as possible of the standard of perfection for that particular breed. A pigeon that could compete with other examples of its breed in a show environment.

Show pen - A specialised pen designed for showing and judging pigeons in a competition or used in a person loft set up for comparing a fancier's own pigeons.

Sire - Father homing pigeon

Smash - The dreaded result of no bird returns from a race.

Squab - A baby pigeon. Usage reserved by fanciers for one that is to be used for food at the table.

Squeaker - A baby pigeon of seven weeks or younger.

Stock bird - Also called a breeder.

Stockings - Feathers on the legs or feet.

Team - The same thing as a kit. (This term is more commonly used in the UK.)

Tick - One or two white feathers located behind the racing homer's eye.

Tumbler - A bird that somersaults in flight.

Trap - A device used for the pigeon's re-entry into the loft after liberation.

Vent bones - Two small bones directly behind and on either side of the breast bone and under its tail.

Wattle - *See beak cere.*

Yearling - A bird in its second year of life.

Young bird - A pigeon that is not a year old.

Record Sheets

You can download these record sheets in A4 size at:
wwww.pigeonsandracingpigeons.com

BREEDING RECORD *for pair number* _____

Date:

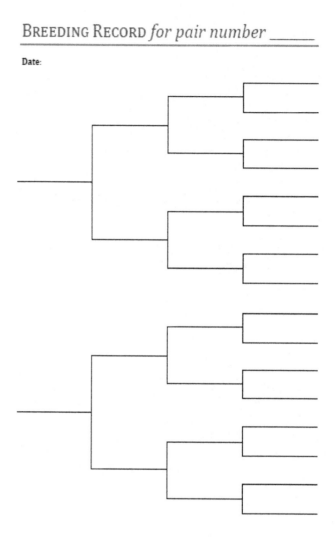

PEDIGREE RECORD *for* _____

Colour: _____ Sex: _____ Year of hatching: _____ Band Number: _____

Sire

Dam

Grand Sire

Grand Dam

Grand Sire

Grand Dam

Great Grand Sire

Great Grand Dam

Great Grand Sire

Great Grand Dam

Great Grand Sire

Great Grand Dam

Great Grand Sire

Great Grand Dam

Name and Address of Breeder:

I declare that the information contained in this Pedigree Record is a correct, factual representation of the named pigeon's ancestry.

Signed: _____ Date: _____

Pigeon Passion

/RACE RESULTS *for date* _____

Date: Distance: Liberation point:

No. of Lofts: No. of Birds:

Weather: Wind-speed: Wind Direction:

Band no.	Position			Condition of the bird on return
	Loft	Club	Combine	

For more information about pigeon keeping
and to buy pigeon keeping products visit
our website at:

www.pigeonsandracingpigeons.com

CPSIA information can be obtained
at www.ICGtesting.com
Printed in the USA
BVOW03s2313170517
484034BV00009B/15/P